Geotechnical Depth Practice Exams
for the Civil PE Exam

Bruce A. Wolle, MSE, PE

PPI®
PPI2PASS.COM
A KAPLAN COMPANY

Report Errors for This Book

PPI is grateful to every reader who notifies us of a possible error. Your feedback allows us to improve the quality and accuracy of our products. Report errata at **ppi2pass.com**.

GEOTECHNICAL DEPTH PRACTICE EXAMS FOR THE CIVIL PE EXAM
First Edition

Current release of this edition: 4

Release History

date	edition number	revision number	update
Jan 2016	1	2	Minor cover updates.
Jan 2018	1	3	Minor corrections. Minor cover updates. Minor formatting and pagination changes.
Oct 2022	1	4	Minor corrections.

Printed in the United States of America.

PPI
ppi2pass.com

ISBN: 978-1-59126-350-0

Table of Contents

Preface

I have enjoyed many opportunities to write practice problems for PPI. I have also taught review courses in which my students have provided generous feedback about their experiences studying for and taking the exam. Therefore, I was excited when PPI proposed *Geotechnical Depth Practice Exams for the Civil PE Exam*. This book gave me the opportunity to apply what I've learned about effective exam review through my writing and teaching.

Most of my students who passed their exams have told me that solving problems is an invaluable way to become familiar with the exam scope and format and to gain confidence retrieving information from both memory and reference materials. And while professional practice and collegiate texts are sources of varied engineering problems, they are no substitute for a collection of exam-like problems. This book presents two realistic practice exams. The problems have been carefully prepared and reviewed to ensure that they cover the exam specifications designated by the National Council of Examiners for Engineering and Surveying (NCEES). Although this book doesn't contain any problems from an actual NCEES exam, its problems closely reflect the content, format, and level of difficulty of the problems encountered on the actual exam. The solutions have been reviewed and edited for both accuracy and thoroughness, so you can check your work and learn from any differences between your solving approach and that presented in this book.

If you find an error or want to propose a more efficient solution method, please submit it for review through PPI's errata reporting website at **ppi2pass.com/errata**. Your comments will be used to improve future editions.

Thanks, and best of luck on your exam!

Bruce A. Wolle, MSE, PE

Acknowledgments

Becoming professionally registered as a civil engineer requires dedication and hard work, two fundamental traits of a successful civil engineer. They are also the traits required to produce a book. This book could not have been possible without the dedication and hard work of a number of people. I'd like to thank my countless past students who, through their comments and questions, showed me how necessary solving exam-like problems is for successful exam review. I appreciate the support and guidance of my many colleagues in my efforts to produce exam review materials that safeguard the standard of care and competent practice of our profession. Like this book, engineering is a collaborative effort.

From PPI, I acknowledge the significant editorial contributions of Sarah Hubbard, director of product development and implementation, and Kate Throckmorton, acquisitions editor. Thank you also to the rest of the product development and implementation staff: Cathy Schrott, production services manager; Magnolia Molcan, editorial project manager; Tyler Hayes, Scott Marley, Heather Turbeville, and Ian A. Walker, copy editors; Kate Hayes, production associate; Tom Bergstrom, technical illustrator; and Scott Miller, EIT, and Alex Valeyev, EIT, calculation checkers.

I would also like to thank Charles Hubbard and Michael Johnson, PhD, PE, for contributing their time and engineering expertise conducting technical reviews of this book.

I am grateful for the support of my wife, Tamara, and my family, friends, and colleagues. Without their support and patience, this book would not have been possible.

Bruce A. Wolle, MSE, PE

Codes and References Used to Prepare This Book

The following codes and references were used to prepare this book. The listed codes have also been adopted as design standards for the geotechnical depth module of the civil PE exam. However, as with engineering practice itself, PE examination is not always based on the most current codes or cutting-edge technology. Similarly, codes, standards, and regulations adopted by state and local agencies often lag issuance by several years. It is likely that the codes that are most current, the codes that you use in practice, and the codes that are the basis of your exam will all be different.

PPI lists on its website the dates and editions of how the codes, standards, and regulations on which NCEES has announced the PE exams are based. It is your responsibility to find out which codes will be tested on your exam.

CODES

ASCE 7: *Minimum Design Loads for Buildings and Other Structures*, 2010, American Society of Civil Engineers, Reston, VA

OSHA: *Safety and Health Regulations for Construction*, 29 CFR Part 1926 (U.S. Federal version), U.S. Department of Labor, Washington, DC

REFERENCES

Bowles, Joseph E. *Foundation Analysis and Design.* New York, NY: McGraw-Hill.

Das, Braja M. and Khaled Sobhan. *Principles of Geotechnical Engineering.* Stamford, CT: Cengage Learning.

Holtz, Robert D., William D. Kovacs, and Thomas C. Sheahan. *An Introduction to Geotechnical Engineering.* Upper Saddle River, NJ: Prentice Hall.

Kramer, Steven L. *Geotechnical Earthquake Engineering.* Upper Saddle River, NJ: Prentice Hall.

Reese, Lymon C., and William F. Van Impe. *Single Piles and Pile Groups Under Lateral Loading.* Boca Raton, FL: CRC Press.

Youd, T. L., I. M. Idriss, R. D. Andrus, et. al. "Liquefaction Resistance of Soils: Summary Report from the 1996 NCEER and 1998 NCEER/NSF Workshops on Evaluation of Liquefaction Resistance of Soils." *Journal of Geotechnical and Geoenvironmental Engineering*, Vol. 127, No. 10. American Society of Civil Engineers.

Introduction

ABOUT THIS BOOK

Geotechnical Depth Practice Exams includes two exams designed to match the format and specifications of the geotechnical depth module of the civil PE exam. Like the actual exam, the exams in this book contain 40 multiple-choice problems, and each problem takes an average of six minutes to solve. Most of the problems are quantitative, requiring calculations to arrive at the correct option. A few are nonquantitative.

Each of the questions will have four answer options, labeled "A," "B," "C," and "D." If the answer options are numerical, they will be displayed in increasing value. One of the answer options is correct (or, will be "most nearly correct"). The remaining answer options are incorrect and may consist of one or more "logical distractors," the term used by NCEES to designate incorrect options that look correct. Incorrect options represent answers found by making common mistakes. These may be simple mathematical errors, such as failing to square a term in an equation, or more serious errors, such as using the wrong equation.

The solutions in this book are presented step-by-step to help you follow the logical development of the solving approach and to provide examples of how you may want to solve similar problems on the exam. Nomenclature is defined to help you quickly identify the variables used and determine appropriate units.

Some solutions include author commentary that uses the following icons for quick identification.

Timesaver: a technique or approach to reduce problem-solving time

Pitfall: a common pitfall or distractor

Solutions presented for each problem may represent only one of several methods for obtaining the correct answer. Alternative problem-solving methods may also produce correct answers.

ABOUT THE EXAM

The Principles and Practice of Engineering (PE) exam is administered by the National Council of Examiners for Engineering and Surveying (NCEES). The civil PE exam is an eight-hour exam divided into a four-hour morning breadth exam and a four-hour afternoon depth exam. The morning breadth exam consists of 40 multiple-choice problems covering eight areas of general civil engineering knowledge: project planning, means and methods, soil mechanics, structural mechanics, hydraulics and hydrology, geometrics, materials, and site development. As the "breadth" designation implies, morning exam problems are general in nature and wide-ranging in scope. All examinees take the same breadth exam.

For the afternoon depth exam, you must select a depth module from one of the five subdisciplines: construction, geotechnical, structural, transportation, or water resources and environmental. The problems on the afternoon depth exam require more specialized knowledge than those on the morning breadth exam. Topics and the distribution of problems on the geotechnical depth module are as follows.

- **Site Characterization (5 questions)**

  Interpretation of existing site data and proposed site development data; subsurface exploration planning; geophysics; drilling techniques; sampling techniques; in situ testing; description and classification of soils; rock classification and characterization; groundwater exploration, sampling, and characterization

- **Soil Mechanics, Laboratory Testing, and Analysis (5 questions)**

  Index properties and testing; strength testing of soil and rock; stress-strain testing of soil and rock; permeability testing properties of soil and rock; effective and total stresses

- **Field Materials Testing, Methods, and Safety (3 questions)**

  Excavation and embankment, borrow source studies, laboratory and field compaction; trench and construction safety; geotechnical instrumentation

- **Earthquake Engineering and Dynamic Loads (2 questions)**

  Liquefaction analysis and mitigation techniques; seismic site characterization, including site classification using ASCE 7; pseudo-static analysis and earthquake loads

- **Earth Structures (4 questions)**

  Slab on grate; ground improvement; geosynthetic applications; slope stability and slope stabilization;

earth dams, levees, and embankments; landfills and caps; pavement structures; settlement

- **Groundwater and Seepage (3 questions)**

  Seepage analysis/groundwater flow; dewatering design, methods, and impact on nearby structures; drainage design/infiltration; grouting and other methods of reducing seepage

- **Problematic Soil and Rock Conditions (3 questions)**

  Karst; collapsible, expansible, and sensitive soils; reactive/corrosive soils; frost susceptibility

- **Earth Retaining Structures (ASD or LRFD) (5 questions)**

  Lateral earth pressure; load distribution; rigid retaining wall stability analysis; flexible retaining wall stability analysis; cofferdams; underpinning; ground anchors, tie-backs, soil nails, and rock anchors for foundations and slopes

- **Shallow Foundations (ASD or LRFD) (5 questions)**

  Bearing capacity; settlement, including vertical stress distribution

- **Deep Foundations (ASD or LRFD) (5 questions)**

  Single-element axial capacity; lateral load and deformation analysis; single-element settlement; downdrag; group effects; installation methods/hammer selection; pile dynamics; pile and drill-shafted load testing; integrity testing methods

All problems on the breadth and depth exams are multiple choice. The problem statement includes all information required to solve the problem, followed by four options. Only one of the four options is correct. Each problem is independent, so incorrectly calculating the answer to one problem will not impact subsequent problems.

For further information and tips on how to prepare for the geotechnical depth module of the civil PE exam, consult the *Civil Engineering Reference Manual* or PPI's civil PE exam FAQs at **ppi2pass.com/cefaq**.

HOW TO USE THIS BOOK

Prior to taking these practice exams, locate and organize relevant resources and materials as if you are taking the actual exam. Refer to the Codes and References section for guidance on recommended materials. All of the codes and references used to prepare this book are also useful resources to study from and take with you to the exam. However, the minimum recommended library for the exam includes the *Civil Engineering Reference Manual.* Before you assemble your library for the exam, visit **ppi2pass.com/stateboards** for a link to your state's board of engineering, and check for any state-specific restrictions on materials you are allowed to bring to the exam. Also, be sure to check NCEES' calculator policy at **ppi2pass.com/calculators** to ensure your calculator can be used on the exam.

The two exams in this book allow you to structure your own exam preparation in the way that is best for you. For example, you might choose to take one exam as a pretest to assess your knowledge and determine the areas in which you need more review, and then take the second after you have completed additional study. Alternatively, you might choose to use one exam as a guide for how to solve different types of problems, reading each problem and solution in kind, and then use the second exam to evaluate what you learned.

Whatever your preferred exam preparation method, these exams will be most useful if you restrict yourself to exam-like conditions when solving the problems. When you are ready to begin an exam, set a timer for four hours. Use only the references you have gathered for use on the exam. Use the space provided near each problem for your calculations, and mark your answer on the answer sheet.

When you finish taking an exam, check your answers against the answer key to assess your performance. Review the solutions to any problems you answered incorrectly or were unable to answer. Read the author commentaries for tips, and compare your problem-solving approaches against those given in the solutions.

Nomenclature

a	acceleration	ft/sec^2
a	area	ft^2
A	area	ft^2
b	width	in
b_r	bond strength	lbf/ft^2
B	unit width, Boussinesq	–
B	width	ft
c	cohesion	lbf/ft^2
c	constant	–
c_A	adhesion	lbf/ft^2
C	circumference	ft
C	correction factor	–
C	Marston's constant for trench fill	–
C_c	compression index	–
C_n	standard penetration test (SPT) correction factor	–
C_P	Marston's constant for broad fill	–
C_Q	factor for normalizing cone penetrometer test (CPT) tip penetration	–
C_u	coefficient of uniformity	–
C_v	coefficient of consolidation	–
C_z	coefficient of curvature	–
CR	compression ratio	–
CRR	cyclic resistance ratio	–
d	depth	ft
d	depth factor, D/H	–
d	diameter	ft
d	distance	ft
D	depth	ft
D	diameter	ft
e	eccentricity	ft
e	void ratio	–
E	modulus of elasticity	lbf/in^2
E_s	Young's modulus	lbf/in^2
ER	energy ratio	–
f	coefficient of friction	–
f	infiltration	in/hr
F	cumulative infiltration	in/hr
F	factor of safety	–
F	normalized friction ratio	–
FS	factor of safety	–
g	gravitational acceleration, 32.2	ft/sec^2
g_c	gravitational constant, 32.2	$lbm\text{-}ft/lbf\text{-}sec^2$
G	universal gravitational constant, 3.320×10^{-11}	$lbf\text{-}ft^2/lbm^2$
h	height	ft
H	height	ft
i	hydraulic gradient	ft/ft
I	moment of inertia	ft^4
I_c	soil behavior type index for CPT	–
k	earth pressure coefficient	–
k	infiltration decay constant	1/hr
k	modulus of subgrade reaction	lbf/in^3
k	relative stiffness	lbf/ft
K	coefficient of permeability	ft/sec
K	earth pressure constant	lbf/ft^3
K	hydraulic conductivity	–
K_c	correction for grain characteristics	–
L	length	ft
LF	bedding load factor	–
LL	liquid limit	–
m	Boussinesq factor	–
m	drainage coefficient	–
m	fraction, x/H	–
m	mass	lbm
m	number of piles along width	–
M	earthquake Richter magnitude	ft-kips
M	mass	lbm
M	moment	ft-kips
n	iteration exponent for liquefaction analysis	–
n	porosity	–
N	bearing capacity factor	–
N	number	–
N	SPT value	blows/ft
N_o	stability number	–
OCR	overconsolidation ratio	–
OM	optimum moisture content	–
p	perimeter	ft
p	pressure	lbf/ft^2

p'	effective pressure	lbf/ft^2
P	force	kips
P	load	kips
P	pressure	lbf/ft^2
P_u	pull-out capacity	kips
PI	plasticity index	–
PL	plastic limit	–
q	bearing capacity	lbf/ft^2
q	compressive strength	lbf/ft^2
q	surcharge	lbf/ft^2
q_c	tip resistance in a cone	lbf/ft^2
q_s	sleeve friction resistance	lbf/ft^2
Q	CPT penetration resistance	–
Q	flow rate	ft^3/sec
Q_s	strength reduction factor	–
r	radius	ft
r	ratio	–
R	force	lbf
R	radius	ft
R	ratio	–
R	resistance	lbf
R	rigidity	–
s	shear strength	lbf/ft^2
s	spacing	in
s	stress	lbf/in^2
S	degree of saturation	%
S	distance per hammer blow	in
S	settlement	ft
S	shape factor	–
S	strength	lbf/ft^2
S	swell	%
S_a	spectral acceleration	ft/sec^2
SG	specific gravity	–
SPT	standard penetration resistance	blows/ft
t	thickness	in
t	time	sec
T	natural period of vibration	sec
T	tensile force	kips
T	thickness	ft
T	transmissivity	–
T_v	time factor	–
u	pore pressure	lbf/ft^2
U	uplift	lbf
U_z	degree of consolidation	–
v	velocity	ft/sec
V	base shear	lbf
V	vertical force surcharge	lbf
V	volume	ft^3
w	load per unit length	lbf/ft
w	water content	–
w	weight	lbf
w	width	ft
W	weight	lbf
W	width	ft
x	distance	ft
x	fraction by weight	–
X	fraction by weight	–
y	aquifer thickness	–
y	vertical displacement	ft
Y	aquifer thickness	–
z	elevation	ft

Symbols

α	adhesion factor	–
α	angle	deg, rad
α	coefficient (liquefaction)	–
β	angle	deg
β	coefficient (liquefaction)	–
β	effective stress factor	–
γ	specific weight	lbf/ft^3
γ	unit weight	lbf/ft^3
δ	friction angle	deg
Δ	deflection	ft
Δ	deformation	ft
Δ	elongation	in
ε	eccentricity	ft
ε	strain	ft/ft
θ	angle	deg, rad
λ	rake angle of retaining wall face	deg
μ	pore pressure	lbf/ft^2
ν	Poisson's ratio	–
ρ	mass density	lbm/ft^3
σ	stress	lbf/ft^2
σ'	effective stress	lbf/ft^2
τ	shear stress	lbf/ft^2
υ	specific volume	ft^3/lbm
υ	strain	in/in
ϕ	angle of internal friction	deg
ϕ'	effective angle of internal friction	deg

Subscripts

α	failure plane
γ	density
0	initial
1N	normalized penetration resistance
a	active, air, allowable, area, or axial
add	additional
app	applied
atm	atmospheric
ave	average
A	adhesion or axial
b	borrow or buoyant
B	borehole
c	cell, cohesive, column, cone tip, consolidation, or correction
col	column
conc	concrete
cont	container
cs	clean sand
d	dead, drainage, or equipotential drop
diff	difference
dr	drainage
D	deviator
e	effective
eff	effective
emb	embankment
E	effective or energy ratio
f	failure, field, fill, final, flow channel, footing, or friction
fs	field sample
g	air (gas)
h	horizontal
i	initial or inside
l	length
L	length or live
LL	liquid limit
max	maximum
mid	midpoint
min	minimum
n	blow count, nominal, normal, or period n
o	at rest, initial, original, out, outlet, outside, or overall
O	outside
OT	overturning
p	bearing, equipotential drop, passive, pile tip, potential, or pressure
q	surcharge
r	radial, radius, ratio, resistance, resultant, or ring
req	required
R	radial, Rankine, ratio, resistance, resisting, resultant, or rod length
s	sample, sand, side friction, skin, soil, solids, specific, or static
sat	saturated
sr	single ring
ss	soil sample
S	sampling method
SL	sliding
t	thickness, time, or total
T	total
u	unconfined, unconfirmed, or undrained
uc	unconfined compression
ult	ultimate
ut	ultimate tensile
v	consolidation, vertical, voids, or volumetric
w	wall, water, or width
x	strong axis or x-direction
y	weak axis, y-direction, or yield
z	at depth z or zero air voids

Practice Exam 1 Instructions

In accordance with the rules established by your state, you may use textbooks, handbooks, bound reference materials, and any approved battery- or solar-powered, silent calculator to work this examination. However, no blank papers, writing tablets, unbound scratch paper, or loose notes are permitted. Sufficient room for scratch work is provided in the Examination Booklet.

You are not permitted to share or exchange materials with other examinees. However, the books and other resources used in this afternoon session do not have to be the same as those used in the morning session.

You will have four hours in which to work this session of the examination. Your score will be determined by the number of questions that you answer correctly. There is a total of 40 questions. All 40 questions must be worked correctly in order to receive full credit on the exam. There are no optional questions. Each question is worth 1 point. The maximum possible score for this section of the examination is 40 points.

Partial credit is not available. No credit will be given for methodology, assumptions, or work written in your Examination Booklet.

Record all of your answers on the Answer Sheet. No credit will be given for answers marked in the Examination Booklet. Mark your answers with the official examination pencil provided to you. Marks must be dark and must completely fill the bubbles. Record only one answer per question. If you mark more than one answer, you will not receive credit for the question. If you change an answer, be sure the old bubble is erased completely; incomplete erasures may be misinterpreted as answers.

If you finish early, check your work and make sure that you have followed all instructions. After checking your answers, you may turn in your Examination Booklet and Answer Sheet and leave the examination room. Once you leave, you will not be permitted to return to work or change your answers.

When permission has been given by your proctor, break the seal on the Examination Booklet. Check that all pages are present and legible. If any part of your Examination Booklet is missing, your proctor will issue you a new Booklet.

WAIT FOR PERMISSION TO BEGIN

Name: ______________________________________
Last First Middle Initial

Examinee number: ______________

Examination Booklet number: ______________

Principles and Practice of Engineering Examination

Afternoon Session
Practice Exam 1

Practice Exam 1 Answer Sheet

Name: ______________________________
Last First Middle Initial

Date: ______________

1. Ⓐ Ⓑ Ⓒ Ⓓ	11. Ⓐ Ⓑ Ⓒ Ⓓ	21. Ⓐ Ⓑ Ⓒ Ⓓ	31. Ⓐ Ⓑ Ⓒ Ⓓ
2. Ⓐ Ⓑ Ⓒ Ⓓ	12. Ⓐ Ⓑ Ⓒ Ⓓ	22. Ⓐ Ⓑ Ⓒ Ⓓ	32. Ⓐ Ⓑ Ⓒ Ⓓ
3. Ⓐ Ⓑ Ⓒ Ⓓ	13. Ⓐ Ⓑ Ⓒ Ⓓ	23. Ⓐ Ⓑ Ⓒ Ⓓ	33. Ⓐ Ⓑ Ⓒ Ⓓ
4. Ⓐ Ⓑ Ⓒ Ⓓ	14. Ⓐ Ⓑ Ⓒ Ⓓ	24. Ⓐ Ⓑ Ⓒ Ⓓ	34. Ⓐ Ⓑ Ⓒ Ⓓ
5. Ⓐ Ⓑ Ⓒ Ⓓ	15. Ⓐ Ⓑ Ⓒ Ⓓ	25. Ⓐ Ⓑ Ⓒ Ⓓ	35. Ⓐ Ⓑ Ⓒ Ⓓ
6. Ⓐ Ⓑ Ⓒ Ⓓ	16. Ⓐ Ⓑ Ⓒ Ⓓ	26. Ⓐ Ⓑ Ⓒ Ⓓ	36. Ⓐ Ⓑ Ⓒ Ⓓ
7. Ⓐ Ⓑ Ⓒ Ⓓ	17. Ⓐ Ⓑ Ⓒ Ⓓ	27. Ⓐ Ⓑ Ⓒ Ⓓ	37. Ⓐ Ⓑ Ⓒ Ⓓ
8. Ⓐ Ⓑ Ⓒ Ⓓ	18. Ⓐ Ⓑ Ⓒ Ⓓ	28. Ⓐ Ⓑ Ⓒ Ⓓ	38. Ⓐ Ⓑ Ⓒ Ⓓ
9. Ⓐ Ⓑ Ⓒ Ⓓ	19. Ⓐ Ⓑ Ⓒ Ⓓ	29. Ⓐ Ⓑ Ⓒ Ⓓ	39. Ⓐ Ⓑ Ⓒ Ⓓ
10. Ⓐ Ⓑ Ⓒ Ⓓ	20. Ⓐ Ⓑ Ⓒ Ⓓ	30. Ⓐ Ⓑ Ⓒ Ⓓ	40. Ⓐ Ⓑ Ⓒ Ⓓ

Practice Exam 1

1. A standard penetration test (SPT) is performed using a split-spoon sampler from a point 14 ft below ground level to create an 8 in diameter borehole. 14 blows are needed to drive the sampler the first 6 in increment, 18 blows are needed for the second 6 in increment, and 21 blows are needed for the third 6 in increment.

correction factor	correction
for an energy ratio of 77%,	
$C_E = 77\%/60\%$	1.28
borehole diameter, C_B	
65–115 mm	1.0
150 mm	1.05
200 mm	1.15
rod length, C_R	
3–4 m	0.75
4–6 m	0.85
6–10 m	0.95
10–30 m	1.0
>30 m	<1.0
sampling method, C_S	
standard sampler	1.0
sampler with liner	1.1–1.3

Using the correction factors in the table, what is most nearly the normalized standard penetration resistance?

(A) 29 blows/ft

(B) 39 blows/ft

(C) 49 blows/ft

(D) 59 blows/ft

2. A modified California ring sampler is used to obtain the in situ moisture content and unit weight of a soil. Five full rings are recovered with a combined soil mass of 961 g. Data pertaining to the samples are given.

average mass of a single ring	44.0 g
average diameter of a single ring	2.42 in
average height of a single ring	1.00 in
moisture content of the sample	14.2%

What is most nearly the dry unit weight of the soil?

(A) 97 lbf/ft^3

(B) 110 lbf/ft^3

(C) 130 lbf/ft^3

(D) 140 lbf/ft^3

3. Atterberg limits and particle size distribution (sieve analysis) obtained from a soil sample are given in the tables.

Atterberg limits	
liquid limit (LL)	plastic limit (PL)
25	18

particle size distribution (U.S. standard sieve, percent passing)			
no. 4 (4.75 mm)	no. 40 (2.00 mm)	no. 100 (0.150 mm)	no. 200 (0.075 mm)
59%	40%	31%	9%

What is the soil classification according to the Unified Soil Classification System (USCS)?

(A) SW

(B) SP

(C) SP-SC

(D) SP-SM

4. An excavation is being performed in an area prone to rockslides. Along the well-defined joints in the wall of the excavation, portions of the rock masses are observed to have slid out. The rock material is fragmented and has a relatively low density. Most likely what type of rock is this?

(A) sedimentary

(B) metamorphic

(C) igneous

(D) not enough information

5. A four-story office building with an approximate plan area of 9000 ft^2 is proposed. The site developer has requested that no improvements to the existing soil conditions will be made. The test boring record of the site is shown, and the building is to have the following live and dead loads.

interior column load	75 kips
exterior wall load	7 kips/ft
floor load	350 lbf/ft^2

depth (ft)	soil classification and remarks	soil symbol	elevation (ft)	sampler type	*N*-count (1st 6 in, 2nd 6 in, 3rd 6 in)
0	brown/black sandy clay (USGS: CL) with surface scattered boulders and talus, very stiff to hard, medium to high moisture, medium to high plasticity		250	SS	2–3–3
5			245	SS	4–5–4
10	brown to dark brown claystone chips and nodules, very stiff to hard, medium moisture to wet, medium plasticity		240	SS	5–6–6
15	dark brown to black claystone with gray/purple streaks, sandy clay auger cuttings, very stiff to hard, medium to high moisture, medium porosity, difficult sampling/auger refused		235	SS	9–50/8 in
				SS	19–50/4 in

Based on the test boring record shown, what type of foundation is structurally and economically most suitable for the building?

(A) mat and pedestal

(B) thickened slab-on-grade

(C) continuous spread footings

(D) drilled cast-in-place concrete piles

6. An earth dam at a fish hatchery exhibits excessive seepage along the downstream face and at the edge of the embankment. To reliably evaluate the seepage condition of the embankment, what are the field observations and soil sampling characteristics that must be obtained?

I. the compression index and coefficient of vertical consolidation

II. the embankment thickness, strata positions, and cohesion

III. the gradation, permeability, and in situ density

IV. the blow counts from the standard penetration test, shear strength, and angle of internal friction

(A) I and III only

(B) I and IV only

(C) II and III only

(D) II and IV only

7. A 2 ft thick pile of demolition debris consisting of bricks and concrete is to be placed over a saturated clay layer for preconsolidation purposes as shown. The water table will be simultaneously lowered to 16 ft below ground surface. The clay has a specific gravity of 2.69, an average water content of 27%, and a compression index of 0.29. Assume the clay layer is normally consolidated and remains saturated after the change in water table height.

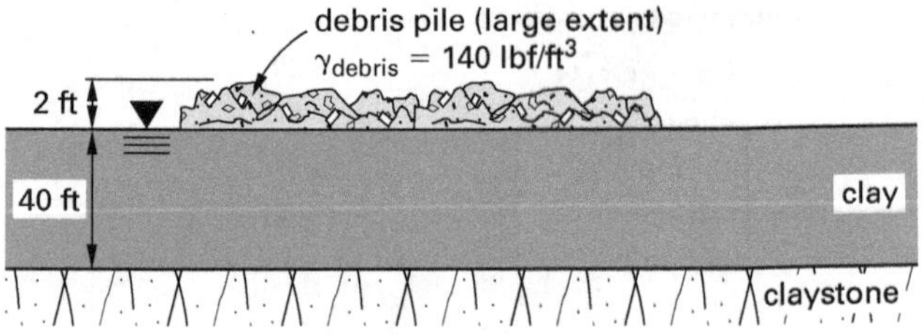

(not to scale)

What is most nearly the settlement of the ground surface?

(A) 2.0 in

(B) 13 in

(C) 25 in

(D) 43 in

8. A clay sample has the given properties.

wet mass	37.59 g
dry mass	24.33 g
specific gravity	2.66
degree of saturation	86%

What is most nearly the void ratio of the clay?

(A) 0.6

(B) 1.5

(C) 1.7

(D) 4.8

9. The in situ dry unit weight of a 1 ft^3 sample of saturated sandy clay is 117.6 lbf/ft^3. The specific gravity of the solids is 2.68, and the liquid limit (LL) is 38. The difference between the water content of the saturated sample and the LL of the soil on a percent solids basis is most nearly

(A) 16%

(B) 22%

(C) 38%

(D) 53%

10. A soil sample with a specific gravity of 2.69 weighs 59 lbf and has a volume of 0.42 ft^3. After the sample is dried, it weighs 55 lbf. What is most nearly the soil's maximum water content without swelling or bleeding?

(A) 7.0%

(B) 10%

(C) 12%

(D) 14%

11. A soil profile is shown.

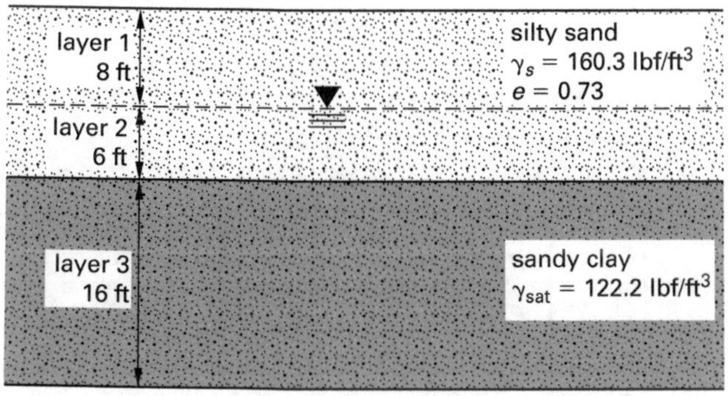

What is most nearly the effective stress at the midpoint of the sandy clay layer?

(A) 870 lbf/ft^2

(B) 980 lbf/ft^2

(C) 1600 lbf/ft^2

(D) 2400 lbf/ft^2

12. A falling-head permeability test is conducted on a 25 cm wide × 25 cm high × 10 cm long soil sample with a standpipe diameter of 4 cm. The water level starts at 187 cm and falls 51 cm in 3 min. What is most nearly the coefficient of permeability of the soil sample?

(A) 3.6×10^{-4} cm/s

(B) 4.5×10^{-4} cm/s

(C) 1.5×10^{-3} cm/s

(D) 1.8×10^{-3} cm/s

13. An unconfined compressive strength test is performed on a typical 6 in wide × 12 in high cylindrical concrete sample. The axial load reaches 54 tons at failure. What is most nearly the shear strength of the concrete?

(A) 1900 lbf/in^2

(B) 2800 lbf/in^2

(C) 3800 lbf/in^2

(D) 7600 lbf/in^2

14. A cohesionless sand is fully drained and tested in a triaxial apparatus. After testing, a well-defined failure plane is measured counterclockwise at an angle of 61° from the horizontal. The deviator stress is 2.60 tons/ft^2 at failure. What is most nearly the effective shear stress on the failure plane?

(A) 0.650 ton/ft^2

(B) 1.10 tons/ft^2

(C) 1.13 tons/ft^2

(D) 2.20 tons/ft^2

15. A constant-head permeability test is performed on the circular soil sample shown. After 4 min, 4 kg of water has been discharged.

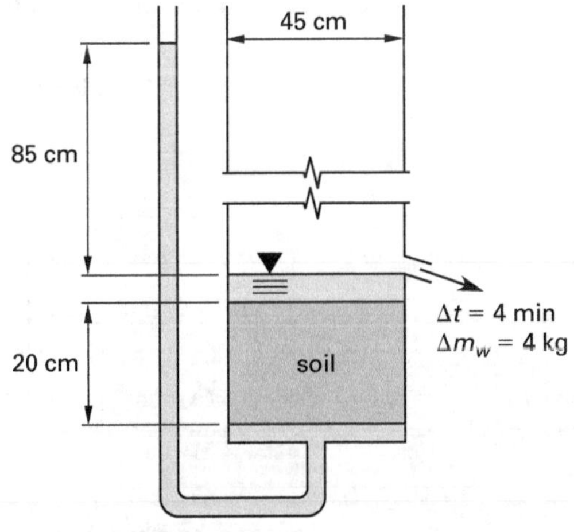

(not to scale)

What is most nearly the coefficient of permeability of the soil sample?

(A) 2.5×10^{-3} cm/s

(B) 8.7×10^{-2} cm/s

(C) 0.15 cm/s

(D) 2.5 cm/s

16. The maintenance facility shown is located next to a reservoir where the water level is proposed to be raised

10 ft. The existing water level elevation is located 22 ft below ground surface at the maintenance facility. Test boring records indicate that the existing subgrade soils consist of 15 ft of compacted clayey sand underlain by silty clay. The clayey sand has an average moisture content of 6.6% and a porosity of 0.28. When dry, 1 ft^3 of the clayey sand weighs 108 lbf in situ.

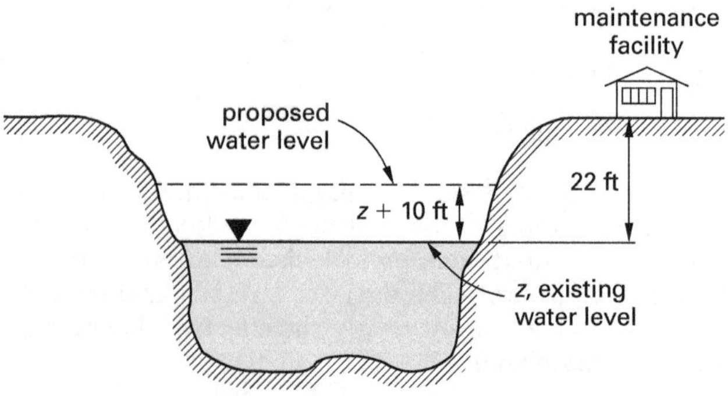

Once the reservoir water level is raised, what will most nearly be the change in effective stress at the base of the clayey sand layer?

(A) 160 lbf/ft^2

(B) 190 lbf/ft^2

(C) 1600 lbf/ft^2

(D) 1800 lbf/ft^2

17. Steady-state vertical seepage occurs in a soil with the profile shown.

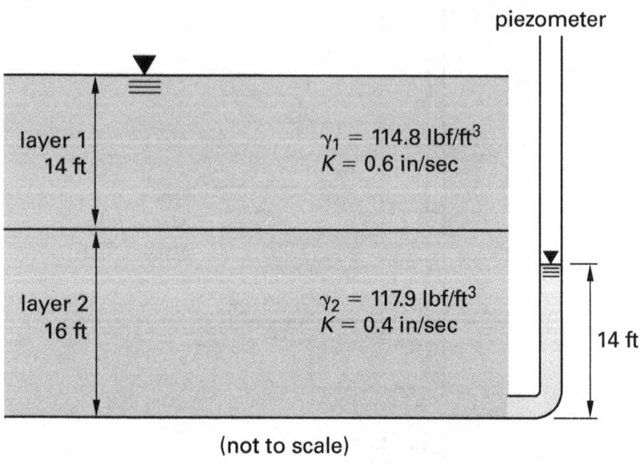

What is most nearly the pore pressure at the base of layer 1?

(A) 380 lbf/ft^2

(B) 510 lbf/ft^2

(C) 870 lbf/ft^2

(D) 990 lbf/ft^2

18. The load-bearing area shown has a soil contact pressure of 6500 lbf/ft^2.

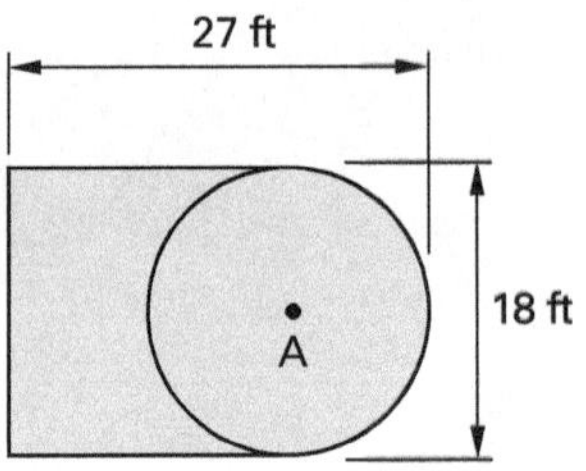

Using Boussinesq contours for pressure distribution, what is most nearly the change in total stress 12 ft below ground surface at point A?

(A) 1100 lbf/ft^2

(B) 2300 lbf/ft^2

(C) 4200 lbf/ft^2

(D) 6800 lbf/ft^2

19. A 30-ton statue is to be placed 9 ft from the edge of an existing retaining wall as shown in the illustration.

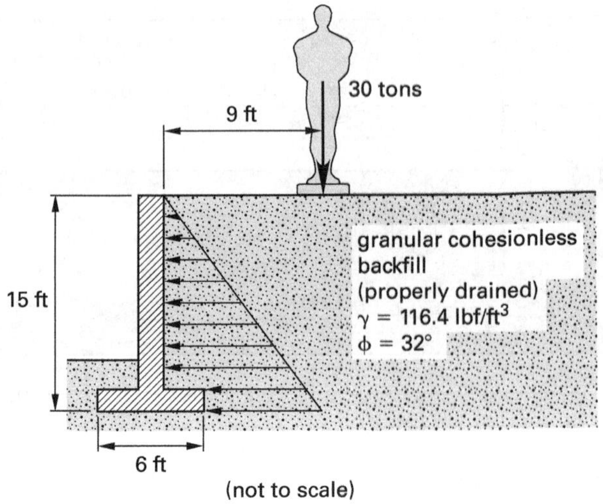

Assuming typical soil elasticity and a Poisson's ratio of 0.5 for the backfill materials, what is most nearly the overturning moment about the toe?

(A) 11 ft-kips

(B) 20 ft-kips

(C) 31 ft-kips

(D) 34 ft-kips

20. A retaining wall has the soil properties shown.

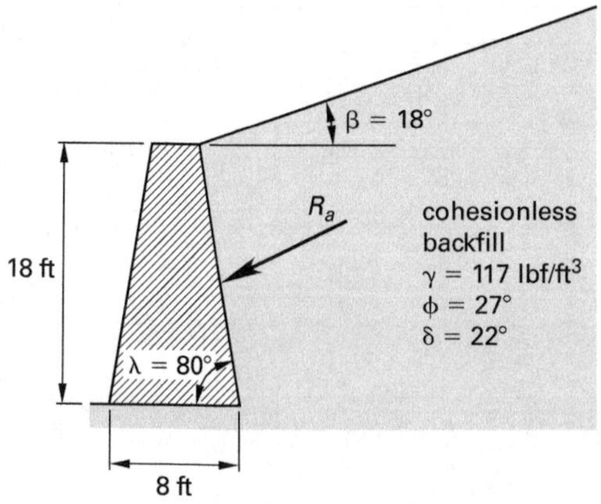

(not to scale)

Using the Coulomb equation, what is most nearly the active resultant, R_a?

(A) 7800 lbf/ft^2

(B) 11,000 lbf/ft^2

(C) 13,000 lbf/ft^2

(D) 15,000 lbf/ft^2

21. What is most nearly the time required for 70% consolidation of the clay layer in the soil profile shown?

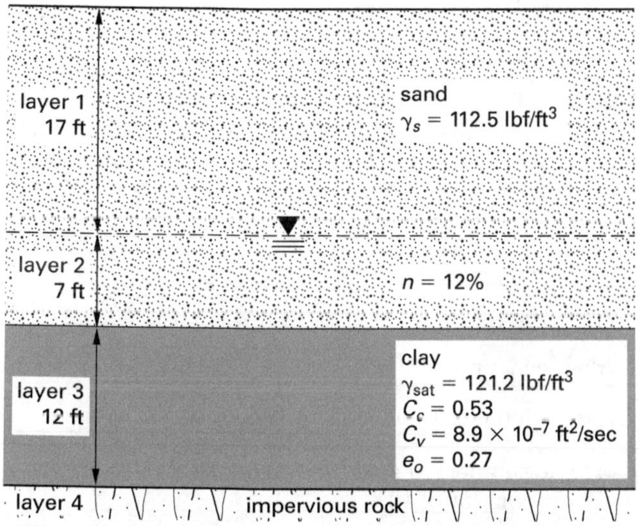

(not to scale)

(A) 62 days

(B) 190 days

(C) 750 days

(D) 890 days

22. Borrow material is proposed as fill for construction of a 250,000 yd^3 embankment. Laboratory test results for the borrow material are shown.

in situ conditions					
moisture content (%)	wet density (lbf/ft^3)	specific gravity	liquid limit	plasticity index	USCS classification
11.5	102.3	2.67	25	9	SC

Specifications require that the embankment fill be compacted to a dry unit weight of 118.8 lbf/ft^3 and ±2% of the optimum moisture content. What is most nearly the quantity of borrow material required to construct the embankment?

(A) 250,000 yd^3

(B) 280,000 yd^3

(C) 290,000 yd^3

(D) 320,000 yd^3

23. A dry soil has a swell potential of 5% when saturated during laboratory testing. The soil is used as a 2 ft thick fill beneath a spread footing foundation where the dry density is 114 lbm/ft^3 after compaction. If a water supply line breaks and floods a portion of the fill, what will be most nearly the increase in fill layer thickness?

(A) 0.01 in

(B) 0.10 in

(C) 1.2 in

(D) 2.4 in

24. A concrete dike retains 60 ft deep water as shown. The coefficient of permeability for the soil is 5×10^{-5} in/sec.

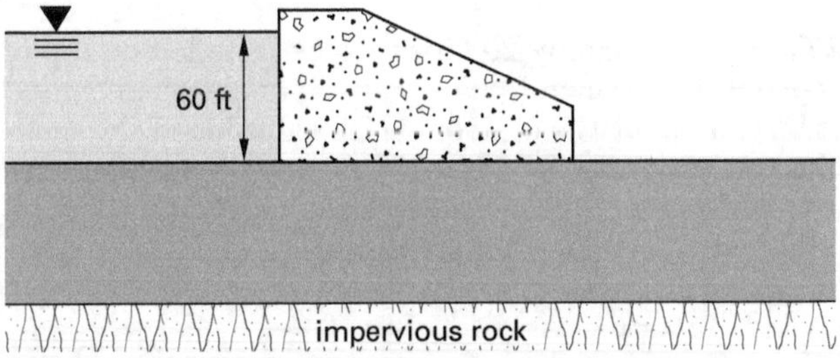

What is most nearly the seepage rate per foot of dike width?

(A) 9.6×10^{-5} ft^2/sec

(B) 6.5×10^{-4} ft^2/sec

(C) 2.5×10^{-4} ft^2/sec

(D) 1.2×10^{-3} ft^2/sec

25. A 30 ft × 30 ft mat foundation with a uniform loading of 2700 lbf/ft² is shown. The normally consolidated clay layer and the groundwater location remain constant.

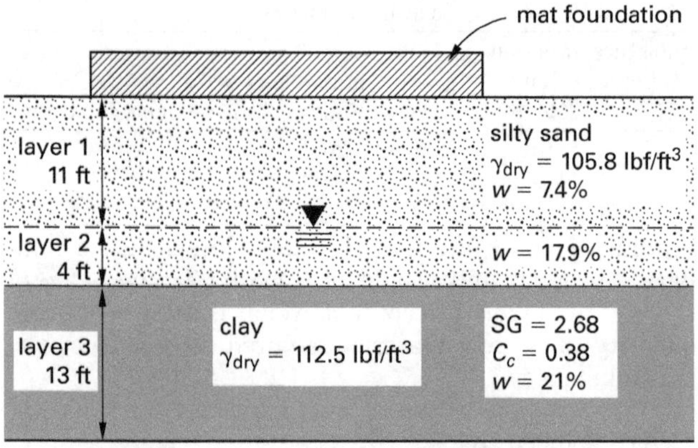

What is most nearly the consolidation settlement of the foundation?

(A) 1 in

(B) 4 in

(C) 7 in

(D) 10 in

26. A water supply tower is located near a fault in an area where the spectral acceleration is estimated at 0.25 ft/sec². The estimated base shear of the tower is 900 lbf. What is most nearly the total weight of the tower?

(A) 30 kips

(B) 120 kips

(C) 230 kips

(D) 1200 kips

27. A sign weighing 20 kips is to be installed on top of two 100 ft long metal poles. For each pole, the moment of inertia about the centroid is 22,000 in⁴, and the modulus of elasticity is 5.9×10^6 lbf/in². Neglecting the weight of each pole, what is most nearly the natural period of vibration of the structure?

(A) 0.10 sec

(B) 0.60 sec

(C) 2.1 sec

(D) 3.0 sec

28. The soil layers shown are liquefiable.

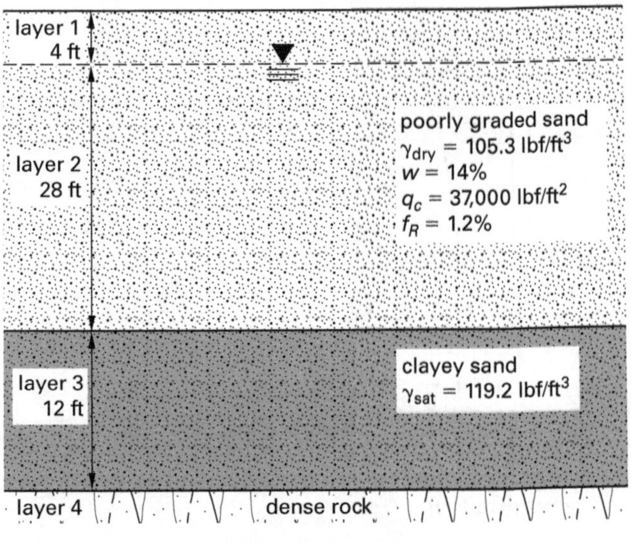

(not to scale)

Using standard atmospheric pressure, what is most nearly the cyclic resistance ratio induced by a moment magnitude 7.5 earthquake ($CRR_{M=7.5}$) 20 ft below the ground surface?

(A) 0.05

(B) 0.07

(C) 0.08

(D) 0.1

29. Installation of a stormwater drainage line consisting of an 18 in diameter flexible pipe at the base of a 6 ft thick embankment is planned. The embankment soil is to be compacted to a unit weight of 115 lbf/ft³. What is most nearly the dead load on the buried pipe?

(A) 260 lbf/ft

(B) 670 lbf/ft

(C) 930 lbf/ft

(D) 1400 lbf/ft

30. The submerged slope shown consists of soils with a moisture content of 12%.

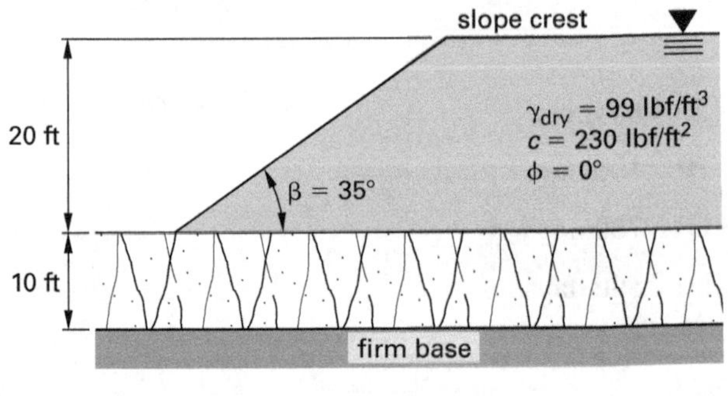

(not to scale)

Using the Taylor method, what is most nearly the cohesive factor of safety of the slope?

(A) 0.60

(B) 1.2

(C) 1.4

(D) 1.9

31. Using the Terzaghi-Meyerhof equation and the Terzaghi bearing capacity factors, what is most nearly the net bearing capacity of the continuous footing shown?

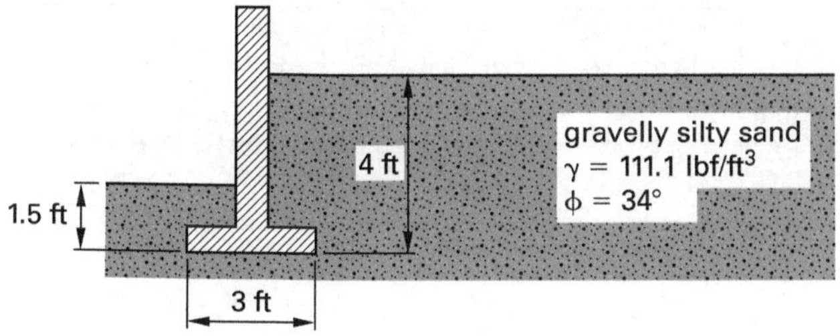

(not to scale)

(A) 9000 lbf/ft^2

(B) 12,000 lbf/ft^2

(C) 16,000 lbf/ft^2

(D) 24,000 lbf/ft^2

32. The earth dam shown retains 75 ft deep water, and is underlain by impervious rock. The coefficient of permeability for the earth dam material is 6×10^{-7} in/sec.

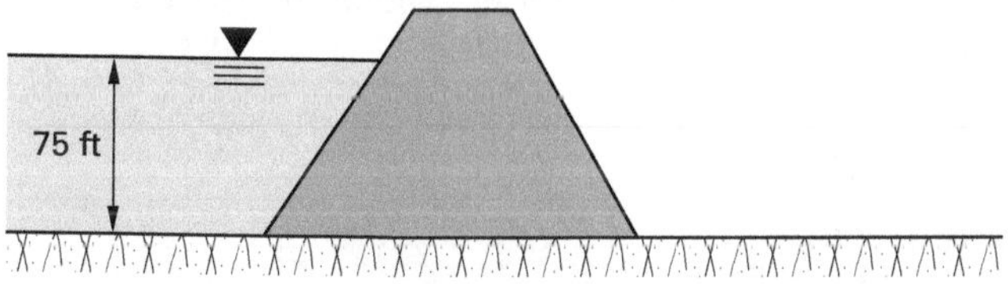

What is most nearly the seepage rate per foot of dam width?

(A) 1.6×10^{-6} ft^2/sec

(B) 2.1×10^{-6} ft^2/sec

(C) 2.5×10^{-6} ft^2/sec

(D) 3.0×10^{-5} ft^2/sec

33. The 30 in diameter concrete pipe buried in the trench as shown has a dead load strength of 750 lbf/ft. Ignore factors of safety associated with the live load.

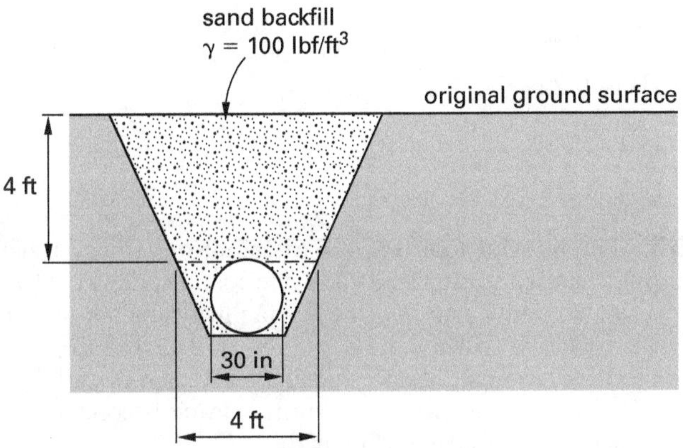

Considering a factor of safety of 1.5 and class C bedding, what is most nearly the maximum live load that can be applied at the top of the pipe?

(A) 530 lbf/ft

(B) 750 lbf/ft

(C) 1300 lbf/ft

(D) 1900 lbf/ft

34. Commencing at a depth of 22 ft in the potentially liquefiable soil profile shown, the results of a corrected (N_{60}) standard penetration test (SPT) are 8, 11, and 11 at 6 in intervals.

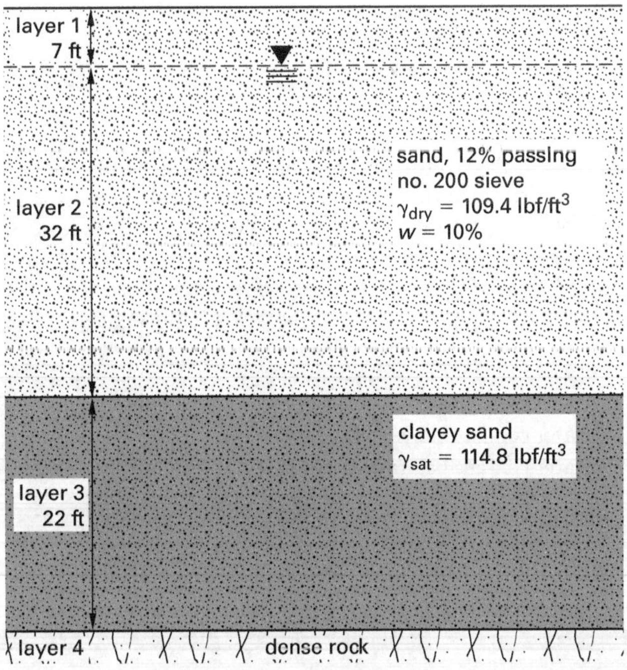

(not to scale)

What is most nearly the cyclic resistance ratio inducible by a moment magnitude 7.5 earthquake ($CRR_{M=7.5}$) for a depth of 23 ft below ground surface?

(A) 0.25

(B) 0.29

(C) 0.30

(D) 0.36

35. The interior footings of a distribution facility must support a structural dead-plus-live load of 120 kips per interior column. The floor of the facility consists of 4 in thick concrete with a unit weight of 145 lbf/ft^3 supported by circular interior footings, set in soil with the profile shown. Assume the column loading is concentric, and neglect the footing self-weight.

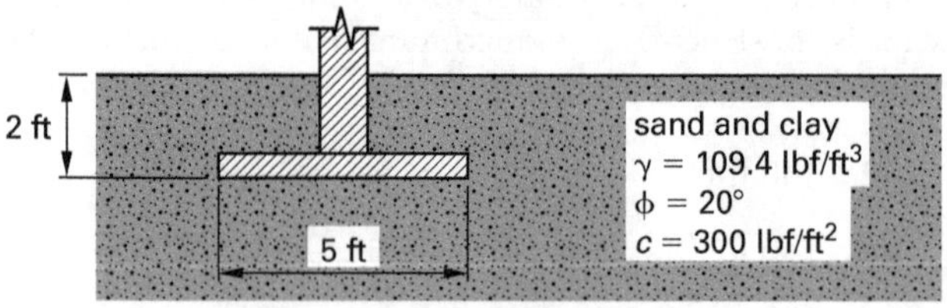

(not to scale)

Using Terzaghi bearing capacity factors, what is most nearly the factor of safety of the given live load with respect to ultimate bearing capacity?

(A) 1.5

(B) 1.9

(C) 2.3

(D) 2.8

36. A 4 ft × 8 ft rectangular footing is constructed 3.5 ft below ground surface. The water table elevation is 2 ft below ground surface, and the soil parameters are given.

dry unit weight	102.3 lbf/ft^3
internal angle of friction	30°
cohesion	150 lbf/ft^2
water content above water table	3%
water content below water table	13%

Using the Terzaghi-Meyerhof equation and Meyerhof bearing capacity factors, what is most nearly the ultimate bearing capacity?

(A) 3810 lbf/ft^2

(B) 11,100 lbf/ft^2

(C) 11,400 lbf/ft^2

(D) 13,200 lbf/ft^2

37. Using the Taylor method, what is most nearly the cohesive factor of safety of the slope shown?

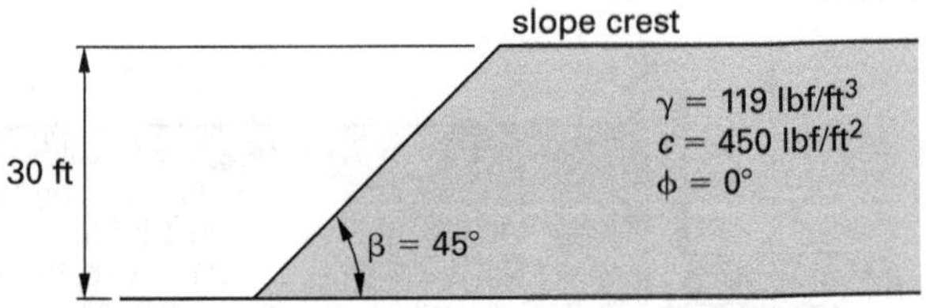

(A) 0.70

(B) 1.2

(C) 1.5

(D) 2.5

38. A sample of sandy clay soil in a 1.34 g container has the given properties.

wet mass plus container	27.2 g
dry mass plus container	21.8 g

What is most nearly the water content of the sample?

(A) 20%

(B) 21%

(C) 25%

(D) 26%

39. What is most nearly the potential Richter magnitude of an earthquake for an active fault length of 6 km?

(A) 3.5

(B) 4.5

(C) 6.4

(D) 7.4

40. A proposed 120 ft × 120 ft raft foundation with below-grade walls will bear 9 ft below ground surface in the soil profile shown. The below-grade walls will bear on the raft foundation. Not including the self-weight of the raft and below-grade wall elements, the superstructure total load is expected to be 1310 lbf/ft^2. Rebound is negligible, and settlement of the dry portion of the clay layer can be ignored.

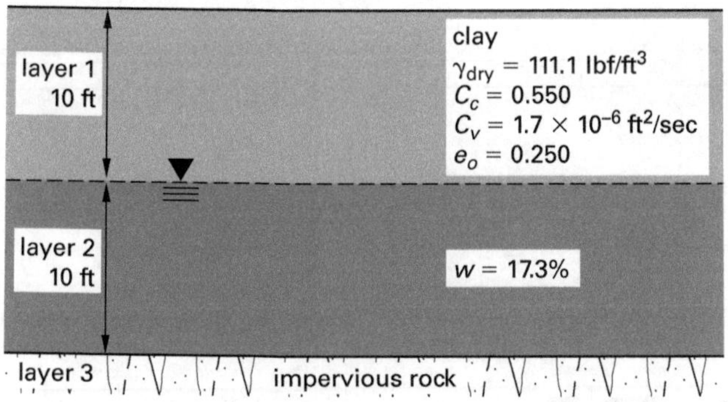

Assuming the unit weight of concrete is 150 lbf/ft^3 and the below-grade walls are 1 ft thick, what is most nearly the expected settlement of the structure and adjoining ground surface?

(A) 2 in

(B) 4 in

(C) 7 in

(D) 9 in

STOP!

DO NOT CONTINUE!

This concludes the Afternoon Session of the examination. If you finish early, check your work and make sure that you have followed all instructions. After checking your answers, you may turn in your examination booklet and answer sheet and leave the examination room. Once you leave, you will not be permitted to return to work or change your answers.

Practice Exam 2 Instructions

In accordance with the rules established by your state, you may use textbooks, handbooks, bound reference materials, and any approved battery- or solar-powered, silent calculator to work this examination. However, no blank papers, writing tablets, unbound scratch paper, or loose notes are permitted. Sufficient room for scratch work is provided in the Examination Booklet.

You are not permitted to share or exchange materials with other examinees. However, the books and other resources used in this afternoon session do not have to be the same as those used in the morning session.

You will have four hours in which to work this session of the examination. Your score will be determined by the number of questions that you answer correctly. There is a total of 40 questions. All 40 questions must be worked correctly in order to receive full credit on the exam. There are no optional questions. Each question is worth 1 point. The maximum possible score for this section of the examination is 40 points.

Partial credit is not available. No credit will be given for methodology, assumptions, or work written in your Examination Booklet.

Record all of your answers on the Answer Sheet. No credit will be given for answers marked in the Examination Booklet. Mark your answers with the official examination pencil provided to you. Marks must be dark and must completely fill the bubbles. Record only one answer per question. If you mark more than one answer, you will not receive credit for the question. If you change an answer, be sure the old bubble is erased completely; incomplete erasures may be misinterpreted as answers.

If you finish early, check your work and make sure that you have followed all instructions. After checking your answers, you may turn in your Examination Booklet and Answer Sheet and leave the examination room. Once you leave, you will not be permitted to return to work or change your answers.

When permission has been given by your proctor, break the seal on the Examination Booklet. Check that all pages are present and legible. If any part of your Examination Booklet is missing, your proctor will issue you a new Booklet.

WAIT FOR PERMISSION TO BEGIN

Name: ______________________________________
Last First Middle Initial

Examinee number: ______________

Examination Booklet number: ______________

Principles and Practice of Engineering Examination

Afternoon Session
Practice Exam 2

Practice Exam 2 Answer Sheet

Name: ______________________________
Last First Middle Initial

Date: ______________

41. (A) (B) (C) (D)	51. (A) (B) (C) (D)	61. (A) (B) (C) (D)	71. (A) (B) (C) (D)
42. (A) (B) (C) (D)	52. (A) (B) (C) (D)	62. (A) (B) (C) (D)	72. (A) (B) (C) (D)
43. (A) (B) (C) (D)	53. (A) (B) (C) (D)	63. (A) (B) (C) (D)	73. (A) (B) (C) (D)
44. (A) (B) (C) (D)	54. (A) (B) (C) (D)	64. (A) (B) (C) (D)	74. (A) (B) (C) (D)
45. (A) (B) (C) (D)	55. (A) (B) (C) (D)	65. (A) (B) (C) (D)	75. (A) (B) (C) (D)
46. (A) (B) (C) (D)	56. (A) (B) (C) (D)	66. (A) (B) (C) (D)	76. (A) (B) (C) (D)
47. (A) (B) (C) (D)	57. (A) (B) (C) (D)	67. (A) (B) (C) (D)	77. (A) (B) (C) (D)
48. (A) (B) (C) (D)	58. (A) (B) (C) (D)	68. (A) (B) (C) (D)	78. (A) (B) (C) (D)
49. (A) (B) (C) (D)	59. (A) (B) (C) (D)	69. (A) (B) (C) (D)	79. (A) (B) (C) (D)
50. (A) (B) (C) (D)	60. (A) (B) (C) (D)	70. (A) (B) (C) (D)	80. (A) (B) (C) (D)

Practice Exam 2

41. 5 ft × 5 ft spread footings are proposed for placement 2.5 ft below ground surface in clean, cohesionless sand. The uncorrected standard penetration test (SPT) result at a depth of 12 ft below ground surface is 12 blows/ft. The unit weight of the sand is 95 lbf/ft^3, and groundwater is not present. Using a factor of safety of 2 and disregarding settlement, what is most nearly the allowable bearing pressure?

(A) 0.70 ton/ft^2

(B) 0.90 ton/ft^2

(C) 1.1 $tons/ft^2$

(D) 1.3 $tons/ft^2$

42. An 18 in diameter steel pile is to be driven to a depth of 40 ft in the soil profile shown.

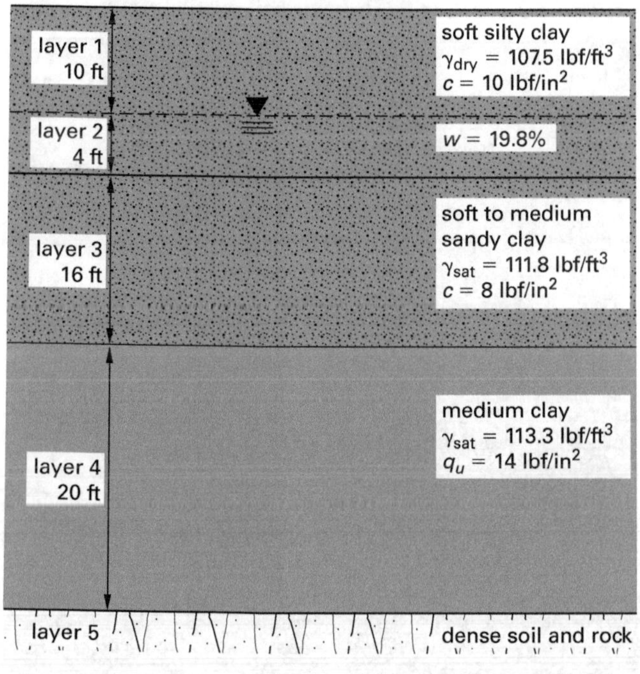

(not to scale)

Using the β-method, what is most nearly the total pile capacity?

(A) 16 kips

(B) 35 kips

(C) 87 kips

(D) 100 kips

43. A 42 in diameter cast-in-place concrete pier is installed in soil with a modulus of subgrade reaction of 140 lbf/in^3. What is most nearly the maximum allowable soil reaction per unit length of pier?

(A) 140 lbf/in

(B) 490 lbf/in

(C) 2900 lbf/in

(D) 5900 lbf/in

44. A shallow continuous (wall) footing is to be constructed as shown in the soil profile. The footing must support a structural load of 30 kips/ft with a factor of safety of 1.5.

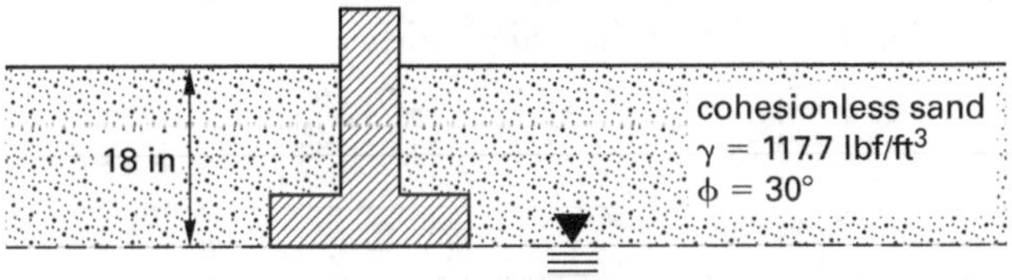

Using Terzaghi-Meyerhof bearing capacity factors and ignoring the footing self-weight and correction for overburden, what is most nearly the required width of the footing?

(A) 1.8 ft

(B) 3.5 ft

(C) 5.3 ft

(D) 7.1 ft

45. Construction of a 35 ft × 35 ft building is proposed in an area with clean, cohesionless sand. The uncorrected standard penetration test (SPT) result at a depth of 20 ft below ground surface is 8 blows/ft. The sand has a unit weight of 103.3 lbf/ft^3, and historic records indicate groundwater could reach 10 ft below ground surface. If a raft foundation must be placed 5 ft below ground surface, what is most nearly the allowable soil pressure?

(A) 0.60 ton/ft^2

(B) 0.90 ton/ft^2

(C) 1.1 tons/ft^2

(D) 1.8 tons/ft^2

46. A building with a 50 ft × 75 ft fully compensated raft foundation and below-grade basement walls is being planned for construction. The required structural support, including the self-weight of the mat, is 2250 tons. The soil at the site is high plasticity clay with a medium-stiff consistency and has the properties given. Groundwater is not present at the site.

dry unit weight	107.9 lbf/ft^3
average water content	12%
unconfined compressive strength	2 lbf/in^2
angle of internal friction	12°

What is most nearly the bearing depth required?

(A) 9.0 ft

(B) 10 ft

(C) 11 ft

(D) 12 ft

47. A 12 in × 12 in precast concrete pile is to be driven to a depth of 35 ft in the soil profile shown.

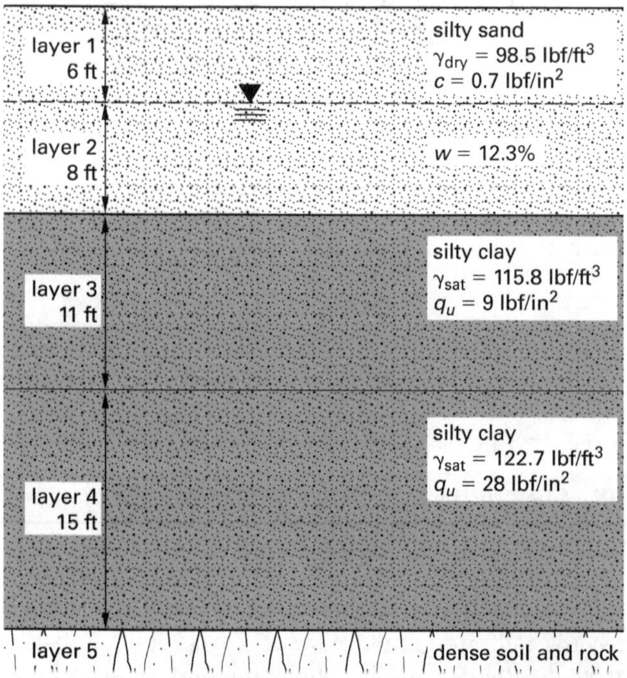

Using the α-method, what is most nearly the total capacity of the pile?

(A) 18 kips

(B) 97 kips

(C) 120 kips

(D) 160 kips

48. A continuous footing is proposed for support of a small perimeter wall in soft, saturated clay. The shear strength of the soil at an elevation proposed for construction of the footing is 150 lbf/ft^2. A factor of safety of 3 is desired, and settlement need not be considered. Using Terzaghi bearing capacity factors, what is most nearly the allowable bearing pressure?

(A) 140 lbf/ft^2

(B) 280 lbf/ft^2

(C) 430 lbf/ft^2

(D) 860 lbf/ft^2

49. A 14 in diameter smooth concrete pile is to be driven to a depth of 28 ft in the soil profile shown.

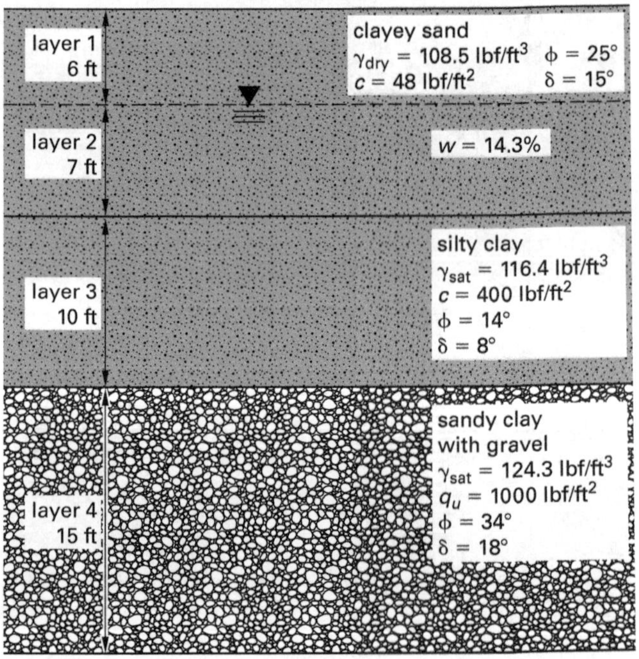

Using the Mohr-Coulomb equation for failure, what is most nearly the skin-friction capacity of the pile?

(A) 32 kips

(B) 46 kips

(C) 140 kips

(D) 190 kips

50. A drop hammer with a striking weight of 10,000 lbf and a fall distance of 36 in is used to drive a closed-tip steel pile to a depth of 30 ft in the soil profile shown. The hammer penetrates 3/4 in per blow.

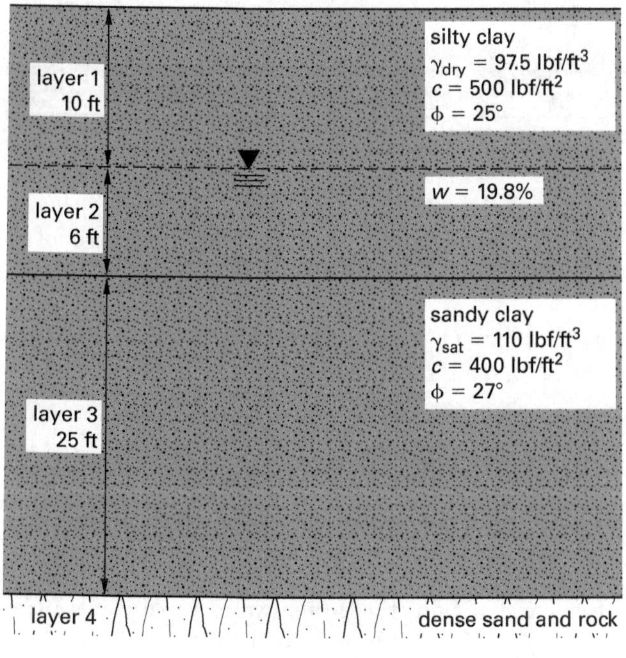

(not to scale)

What is most nearly the allowable pile capacity?

(A) 34 kips

(B) 65 kips

(C) 100 kips

(D) 150 kips

51. A shallow circular footing is to be constructed in the soil profile shown. The footing must support a structural load of 28 kips with a factor of safety of 2.5.

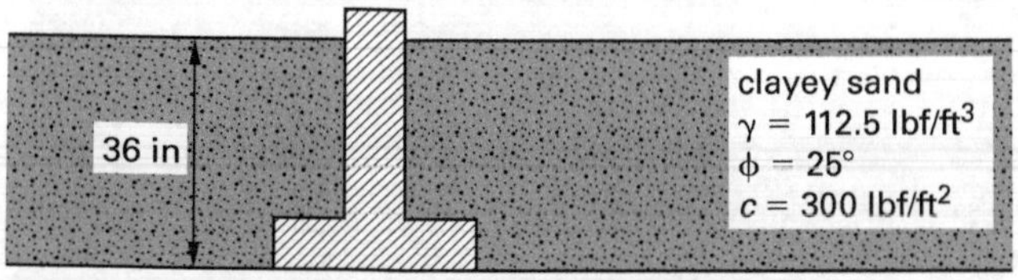

Using Terzaghi bearing capacity factors and ignoring the footing self-weight and correction for overburden, what is most nearly the required footing diameter?

(A) 1.5 ft

(B) 2.0 ft

(C) 2.5 ft

(D) 3.0 ft

52. Young's modulus data for various depths of an 18 in diameter precast concrete pile are given in the table.

depth, D (ft)	Young's modulus, E (kips/ft^2)
5	600
10	1200
15	1800
20	2400
25	3000

Assuming a unit lateral deflection of 1 in, what is most nearly the soil reaction of the pile at a depth of 15 ft?

(A) 300 lbf/in

(B) 600 lbf/in

(C) 1300 lbf/in

(D) 2400 lbf/in

53. Based on a driven pile's soil reaction to vertical depth (p-y) curve, the estimated maximum soil reaction at a point 8 ft below ground surface is 650 lbf/in. If the actual soil reaction at this point is 300 lbf/in, what is most nearly the maximum soil reaction?

(A) 0.50%

(B) 10%

(C) 26%

(D) 46%

54. A 3 by 4 pile group consisting of 30 in diameter piles spaced 78 in on center extends 20 ft into the soil profile shown.

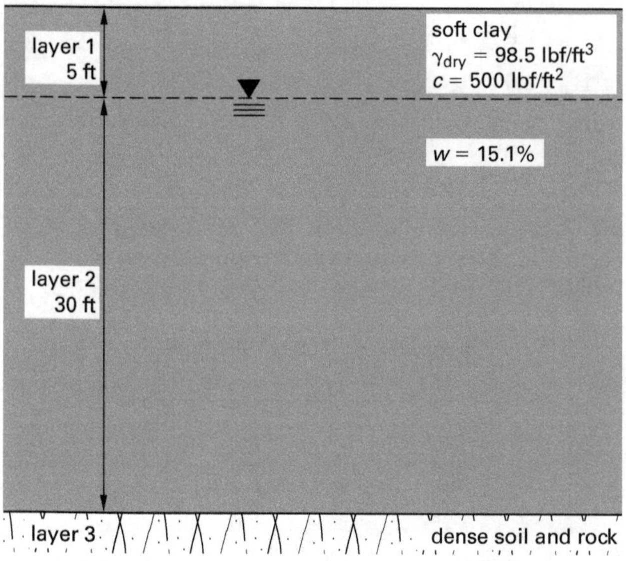

(not to scale)

Using the α-method and disregarding end bearing, the self-weight of the piles, and the weight of the pile cap, what is most nearly the total skin-friction capacity of the pile group?

(A) 250 kips

(B) 750 kips

(C) 850 kips

(D) 940 kips

55. Two 30 in diameter piles are spaced 42 in on center. If the structural loading, length of embedment, and soil conditions are the same for both piles, will the piles act individually or as a group?

(A) The piles will act individually.

(B) The piles will act as a group.

(C) Unable to determine due to lack of soil profile information.

(D) Unable to determine due to lack of pile dimension information.

56. The gravity wall shown consists of concrete with a unit weight of 145 lbf/ft^3 and has a key located a negligible distance from the heel.

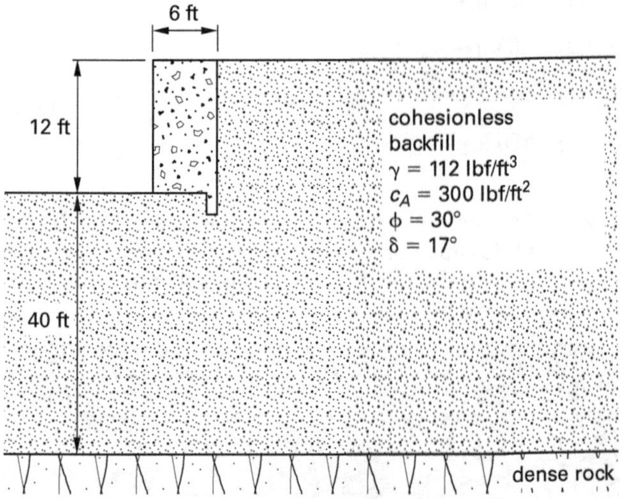

(not to scale)

What is most nearly the resistance to sliding per foot of wall length?

(A) 450 lbf/ft

(B) 2700 lbf/ft

(C) 6000 lbf/ft

(D) 7800 lbf/ft

57. A 4 by 4 pile group consisting of 30 in diameter piles spaced 90 in on center and extending 27 ft below ground surface is constructed in the soil profile shown. After construction, the groundwater table elevation is raised 10 ft. The total load used for evaluating settlement of the pile group is to be calculated as 30% of the ultimate capacity. The ultimate capacity is to be calculated based on individual pile capacity using the α-method for friction capacity and end bearing. Disregard pile group self-weight.

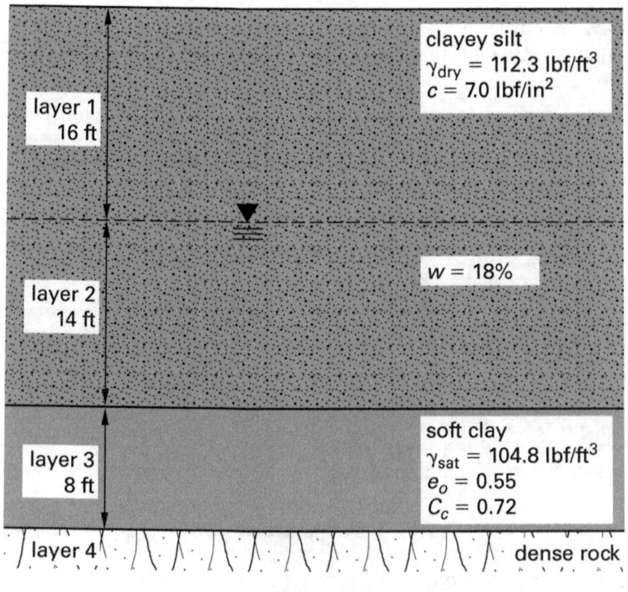

(not to scale)

What is most nearly the settlement of the piles?

(A) 0.5 in

(B) 1 in

(C) 5 in

(D) 8 in

58. Pile loading and unloading reactions are given in the table. To meet design requirements, the maximum allowable pile load is 75% of the test load that results in a settlement ratio of not more than 0.003 in/kip.

settlement (in)	loading (kips)	unloading (kips)
0.00	0	–
0.25	50	–
0.50	100	–
0.75	150	–
1.00	200	0
1.25	290	50
1.50	360	200
1.75	430	350
2.00	500	500

What is most nearly the maximum allowable pile load?

(A) 180 kips

(B) 270 kips

(C) 360 kips

(D) 500 kips

59. An excavation is to extend 10 ft below ground surface in the soft clayey soil as shown.

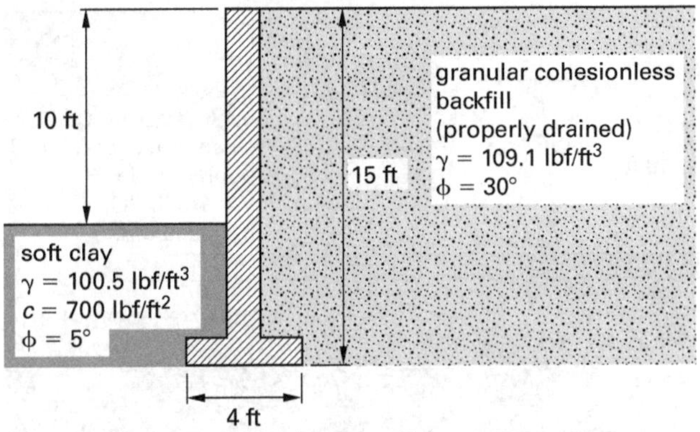

(not to scale)

What is most nearly the passive earth pressure development per foot of footing width beneath the base of the excavation?

(A) 600 lbf/ft²

(B) 1200 lbf/ft²

(C) 1600 lbf/ft²

(D) 2100 lbf/ft²

60. A 2 by 3 pile group consisting of 20 in diameter piles spaced 60 in on center extends 21 ft into the soil profile shown. The load on the group is 180 kips.

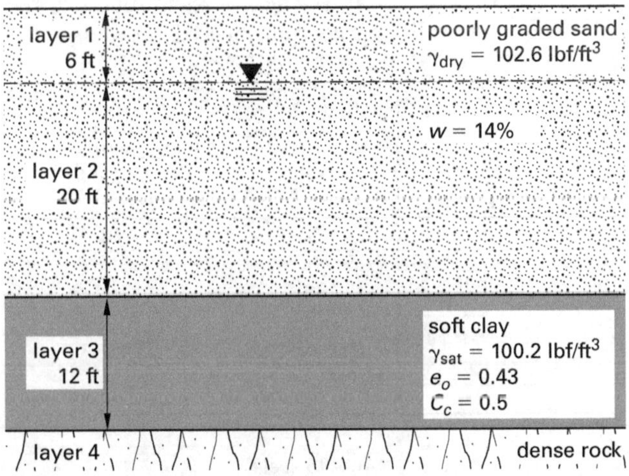

(not to scale)

If the piles act as a group, what is most nearly the total expected settlement of the piles?

(A) 0.5 in

(B) 2 in

(C) 5 in

(D) 8 in

61. The keyless gravity wall shown consists of concrete with a unit weight of 145 lbf/ft^3.

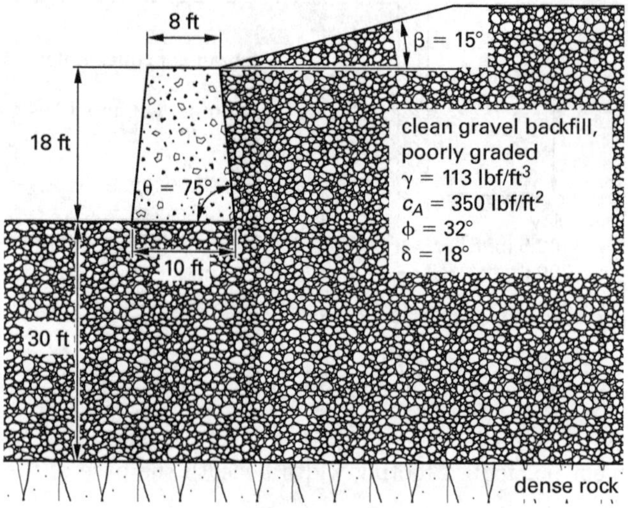

(not to scale)

What is most nearly the factor of safety against sliding?

(A) 0.40

(B) 1.5

(C) 2.3

(D) 3.5

62. What is most nearly the active resultant force per unit length of wall for the cantilevered retaining wall shown?

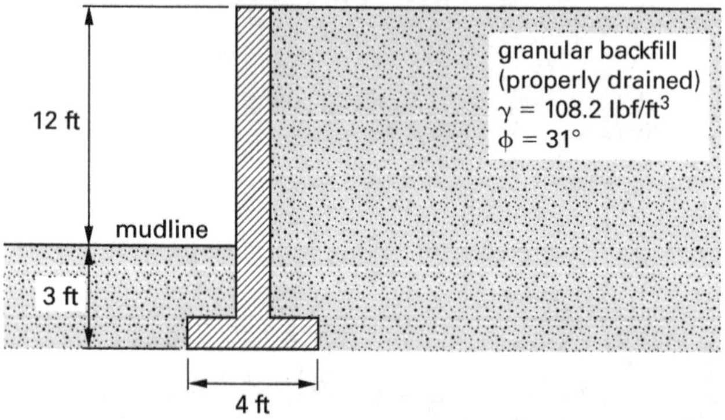

(not to scale)

(A) 260 lbf/ft

(B) 1600 lbf/ft

(C) 3900 lbf/ft

(D) 12,000 lbf/ft

63. A cantilevered retaining wall consisting of concrete with a unit weight of 145 lbf/ft^3 is shown.

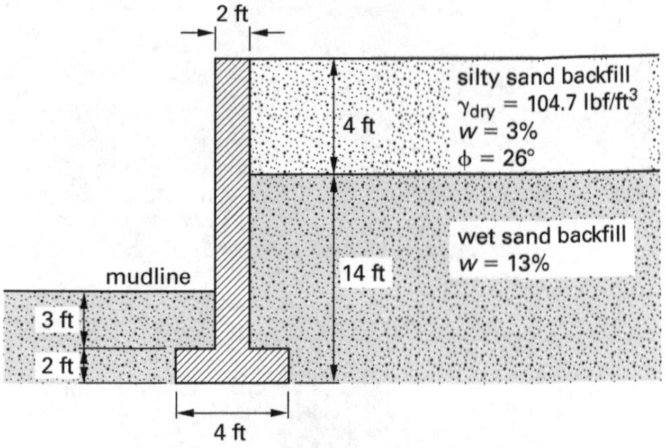

(not to scale)

Ignoring passive earth pressure, what is most nearly the factor of safety against overturning?

(A) 0.25

(B) 0.42

(C) 1.5

(D) 3.0

64. The mechanically reinforced retaining wall shown is to be stabilized using 1.5 in diameter metal rods. The rods will be spaced at 2.5 ft horizontal intervals on center and at 3 ft vertical intervals commencing at a depth of 1.5 ft below the backfill surface.

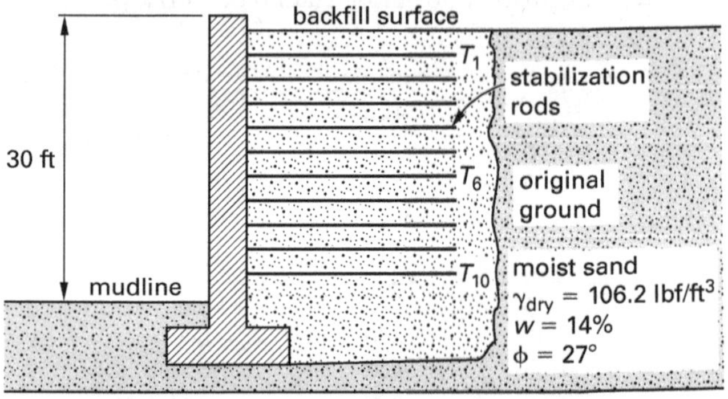

If passive resistance is negligible, and the friction angle between the metal and backfill is 16°, what is most nearly the tension in stabilizing rod 6, T_6?

(A) 5.0 kips

(B) 5.7 kips

(C) 6.5 kips

(D) 7.0 kips

65. A 30 ft deep trench is to be excavated in the soil profile shown.

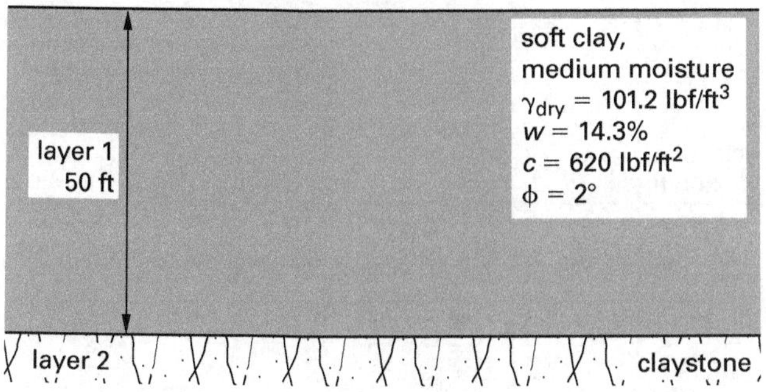

(not to scale)

What is most nearly the width of the excavation required to maintain a minimum factor of safety of 1.3 against bottom heave?

(A) 5.0 ft

(B) 10 ft

(C) 15 ft

(D) 20 ft

66. A 25 ft deep excavation in the soil profile shown is to be braced using an anchored flexible bulkhead installed 40 ft deep and supported with a tie rod. The bulkhead is fixed from rotation and horizontal movement. The tie rod will be anchored behind the active wedge at a depth of 6 ft below ground surface.

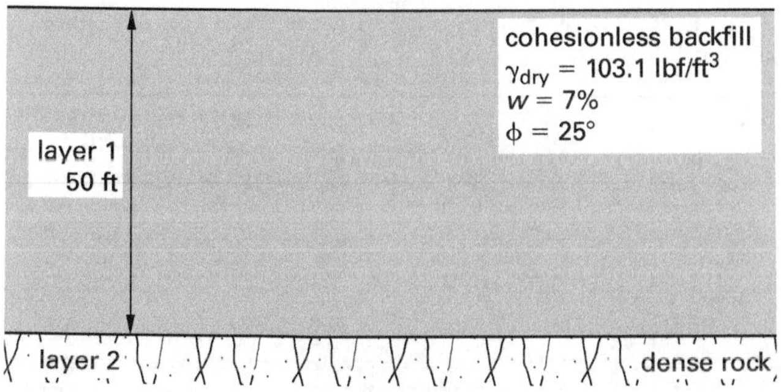

(not to scale)

What is most nearly the tensile force in the tie rod per unit length of bulkhead?

(A) 3400 lbf/ft

(B) 7400 lbf/ft

(C) 8900 lbf/ft

(D) 9500 lbf/ft

67. A 36 ft deep braced excavation is to be completed in the soil profile shown. Struts are proposed for every 6 ft of depth.

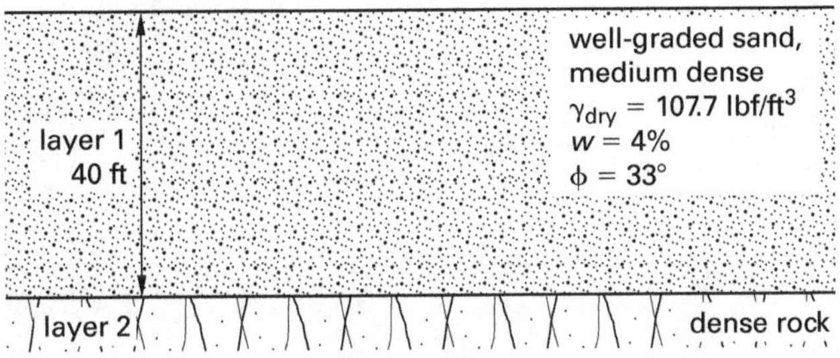

(not to scale)

Using the Tschebotarioff trapezoidal earth pressure distribution and evaluating at the midpoint of the excavation depth, what is most nearly the expected maximum lateral load on each strut per unit width of bracing?

(A) 940 lbf/ft

(B) 1900 lbf/ft

(C) 5600 lbf/ft

(D) 8400 lbf/ft

68. The mechanically reinforced retaining wall shown is to be stabilized using 2.5 in wide × 3/8 in thick rectangular reinforcing strips spaced at 3 ft horizontal intervals on center and at 2.5 ft vertical intervals on center commencing at 1.25 ft below the backfill surface. Passive resistance and the strip thickness contribution to frictional resistance are negligible.

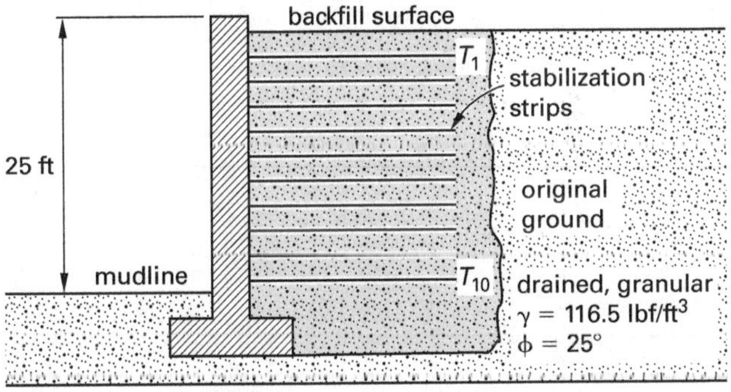

Most nearly, how long must the reinforcing strips be if the friction angle between the metal and the soil is 18°?

(A) 16 ft

(B) 23 ft

(C) 30 ft

(D) 39 ft

69. What is most nearly the thickness of the aquifer for the soil profile shown?

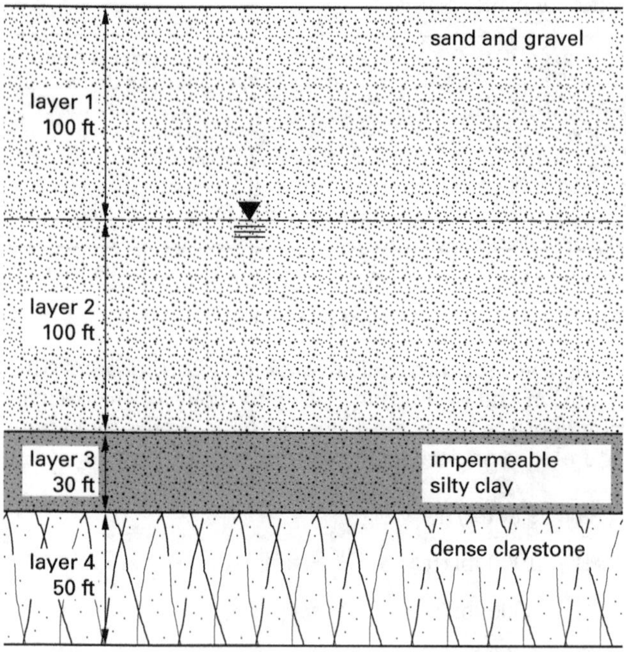

(not to scale)

(A) 100 ft

(B) 200 ft

(C) 230 ft

(D) 280 ft

70. In the aquifer profile shown, what kind of zone is layer 2?

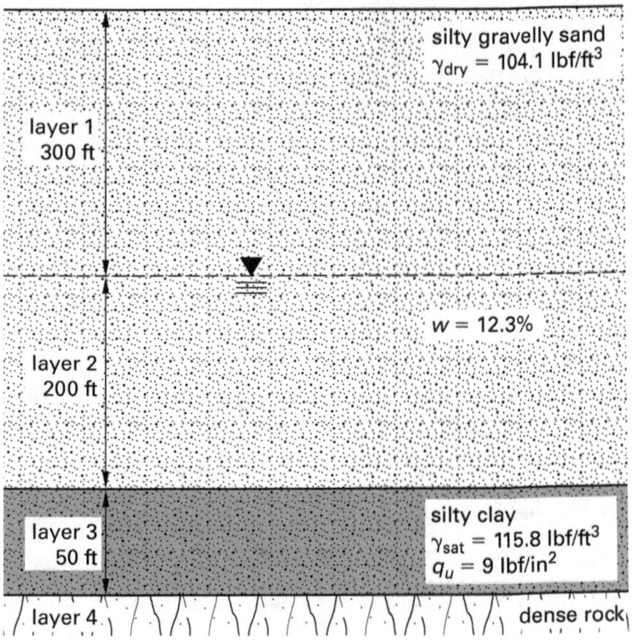

(not to scale)

(A) artesian zone

(B) confined zone

(C) phreatic zone

(D) vadose zone

71. The groundwater in an aquifer with the soil profile shown is moving at a rate of 3 mi every 180 days.

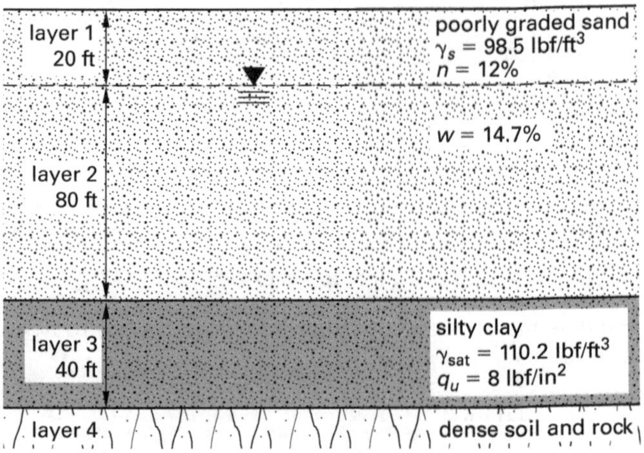

(not to scale)

If a contaminant is introduced into the aquifer, what will be most nearly its pore velocity?

(A) 5.7×10^{-3} in/sec

(B) 1.7×10^{-2} in/sec

(C) 0.10 in/sec

(D) 0.20 in/sec

72. Groundwater flows along the length of a 50 mi wide × 100 mi long aquifer with the soil profile shown. The elevation change along the length of the aquifer is 200 ft.

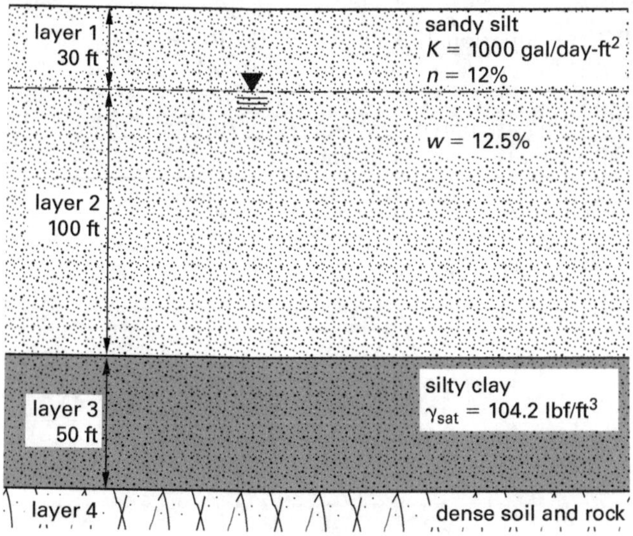

(not to scale)

What is most nearly the overall groundwater flow rate?

(A) 3.6 ft³/sec

(B) 15 ft³/sec

(C) 13,000 ft³/sec

(D) 38,000 ft³/sec

73. A well has a cone of depression with a 700 ft radius of influence. An 18 in diameter well is proposed for an unconfined aquifer nearby. The aquifer phreatic surface is 100 ft below ground surface, and the depth to impermeable rock is 500 ft below ground surface. The new well is designed to supply water at a rate of 500 gal/min, will have a screen located 200 ft below ground surface, and will penetrate the full depth of the aquifer down to the impermeable layer. The edge of the new well's cone of depression must be at least 500 ft from the edge of the existing well's cone of depression. The flow will be maintained such that drawdown will not extend below the top of the screen, and that the transmissivity is 8000 gal/day-ft. What is most nearly the minimum allowable distance between the wells?

(A) 770 ft

(B) 1200 ft

(C) 1500 ft

(D) 1900 ft

74. The water level in a 20 in diameter well is located 75 ft below ground surface within a low permeability zone. The bottom of the low permeability zone is 125 ft below ground surface. The well penetrates 200 ft below ground surface to impermeable rock. The installed pump provides a maximum drawdown of 30 ft and a radius of influence of 1000 ft. The hydraulic conductivity of the water in the well is 100 gal/day-ft². What is most nearly the safe yield available from the well?

(A) 110,000 gal/day

(B) 200,000 gal/day

(C) 220,000 gal/day

(D) 240,000 gal/day

75. A 100,000 ft² site with 4 ft deep water (measured up from ground level) is to be drained for construction. The proposed subgrade drainage system consists of wells that are designed to draw the water from 12 ft below the water surface. The hydraulic conductivity of the subsurface soil is 5×10^{-4} in/sec, and flow is downward in one dimension. What is most nearly the initial flow rate required?

(A) 0.70 ft³/sec

(B) 1.4 ft³/sec

(C) 2.1 ft³/sec

(D) 2.8 ft³/sec

76. A soil consists of poorly graded silty sand with the given properties.

hydraulic conductivity, K	2.5×10^{-4} in/sec
initial infiltration capacity, f_0	1.8 in/hr
equilibrium capacity, f_c	0.16 in/hr
infiltration decay constant, k	0.6 1/hr

If the soil becomes saturated during a 2 hr rainstorm, what is most nearly the quantity by which the surface infiltration will exceed the hydraulic conductivity of the subsurface soils?

(A) 0.60×10^{-4} in/sec

(B) 2.5×10^{-4} in/sec

(C) 3.1×10^{-4} in/sec

(D) 5.0×10^{-4} in/sec

77. A shoring system with tiebacks is being installed. Each tieback will be supported by a 4 in diameter hardened-steel rock bolt with a tensile capacity of 200 kips, embedded to a depth of 8 ft into rock. Grout will be used to form the bonds. Ignoring the tensile contribution at the end of the bolt and using a factor of safety of 3, what is most nearly the bond strength between the grout and rock needed to prevent pullout?

(A) 170 lbf/in^2

(B) 200 lbf/in^2

(C) 330 lbf/in^2

(D) 500 lbf/in^2

78. Soldier piles and lagging will support an excavation in the soil profile shown. The horizontal spacing between the piles is 10 ft on center, and the maximum allowable moment on the lagging is 3600 ft-lbf.

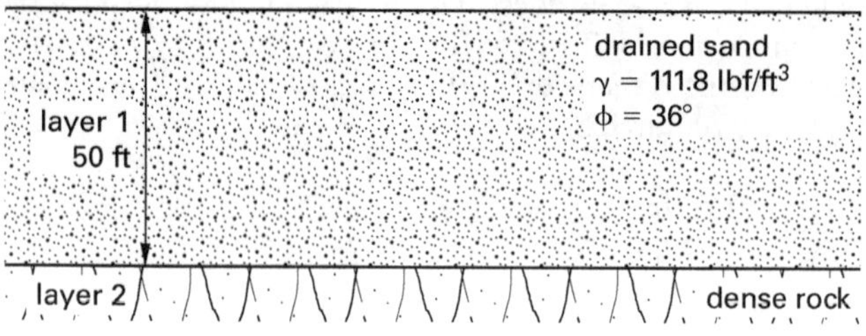

(not to scale)

If the pressure distribution on the horizontal lagging is rectangular, what is most nearly the maximum excavation depth allowed?

(A) 8.0 ft

(B) 15 ft

(C) 25 ft

(D) 50 ft

79. A crane is being used for a construction project in Occupational Safety and Health Administration (OSHA) type A soil. Each outrigger pad is 81 ft^2 and experiences a maximum load of approximately 100 tons. Using the OSHA allowable bearing capacity for type A soil, what is most nearly the factor of safety with respect to soil failure?

(A) 0.40

(B) 0.80

(C) 1.2

(D) 1.6

80. Occupational Safety and Health Administration (OSHA) type C soil is excavated to a depth of 12 ft at a slope of $1\frac{1}{2}$:1. After excavation, tension cracks and sloughing are noted near the crest. To what slope should the excavation be reconstructed?

(A) $\frac{3}{4}$:1

(B) 1:1

(C) $1\frac{1}{2}$:1

(D) 3:1

STOP!

DO NOT CONTINUE!

This concludes the Afternoon Session of the examination. If you finish early, check your work and make sure that you have followed all instructions. After checking your answers, you may turn in your examination booklet and answer sheet and leave the examination room. Once you leave, you will not be permitted to return to work or change your answers.

Answer Keys

Practice Exam 1 Answer Key

Question	Answer
1.	C
2.	B
3.	C
4.	A
5.	D
6.	C
7.	C
8.	C
9.	B
10.	B
11.	C
12.	A
13.	A
14.	B
15.	A
16.	A
17.	B
18.	C
19.	C
20.	B
21.	C
22.	D
23.	C
24.	A
25.	D
26.	B
27.	C
28.	C
29.	D
30.	C
31.	B
32.	C
33.	A
34.	C
35.	A
36.	C
37.	A
38.	D
39.	C
40.	C

Practice Exam 2 Answer Key

Question	Answer
41.	D
42.	D
43.	D
44.	D
45.	C
46.	B
47.	B
48.	B
49.	B
50.	A
51.	C
52.	C
53.	D
54.	B
55.	B
56.	D
57.	B
58.	B
59.	D
60.	B
61.	C
62.	C
63.	B
64.	B
65.	B
66.	C
67.	C
68.	D
69.	B
70.	C
71.	C
72.	B
73.	C
74.	B
75.	C
76.	A
77.	D
78.	B
79.	C
80.	D

Solutions
Practice Exam 1

1. The standard penetration resistance, or N-value, is obtained by counting the number of blows needed to drive the sampler 12 in after the initial 6 in seating drive. The blow counts for the final two 6 in increments are added together to calculate the N-value in blows per foot.

$$\begin{aligned} N &= N_{\text{2nd 6 in}} + N_{\text{3rd 6 in}} = 18 \text{ blows} + 21 \text{ blows} \\ &= 39 \text{ blows/ft} \end{aligned}$$

The N_{60} value is the penetration resistance normalized for the given hammer efficiency (energy ratio). Apply the correction factors as given in the table. The rod length is

$$14 \text{ ft} + \frac{(3)(6 \text{ in})}{12 \ \frac{\text{in}}{\text{ft}}} = 15.5 \text{ ft}$$

C_R for a 15.5 ft (4.72 m) rod is 0.85. C_B for an 8 in (203 mm) diameter borehole is 1.15.

$$\begin{aligned} (N_1)_{60} &= NC_EC_BC_RC_S \\ &= \left(39 \ \frac{\text{blows}}{\text{ft}}\right)\left(\frac{0.77}{0.60}\right)(1.15)(0.85)(1.0) \\ &= 48.9 \text{ blows/ft} \quad (49 \text{ blows/ft}) \end{aligned}$$

Author Commentary

This problem shows the importance of recognizing and using correction factors when they are given in the problem statement data.

The answer is (C).

2. The total mass of the soil sample can be calculated as the mass of the field sample minus the total mass of the rings.

$$\begin{aligned} m_t &= m_{\text{fs}} - m_{r,t} \\ &= 961 \text{ g} - (5 \text{ rings})\left(44.0 \ \frac{\text{g}}{\text{ring}}\right) \\ &= 741 \text{ g} \end{aligned}$$

Calculate the total unit weight of the recovered soil sample and convert it to dry unit weight.

The volume of a single ring is

$$\begin{aligned} V_r &= \frac{\pi}{4}D^2H \\ &= \frac{\pi}{4}\left((2.42 \text{ in})\left(2.54 \ \frac{\text{cm}}{\text{in}}\right)\right)^2\left((1.00 \text{ in})\left(2.54 \ \frac{\text{cm}}{\text{in}}\right)\right) \\ &= 75.4 \text{ cm}^3 \end{aligned}$$

Converting to customary U.S. units, the total unit weight of the soil sample is

$$\begin{aligned} \gamma_t &= \frac{m_t}{V_t} = \left(\frac{741 \text{ g}}{(5 \text{ rings})\left(75.4 \ \frac{\text{cm}^3}{\text{ring}}\right)}\right)\left(62.4 \ \frac{\frac{\text{lbf}}{\text{ft}^3}}{\frac{\text{g}}{\text{cm}^3}}\right) \\ &= 122.7 \text{ lbf/ft}^3 \end{aligned}$$

The dry unit weight of the soil is

$$\begin{aligned} \gamma_{\text{dry}} &= \frac{\gamma_t}{1+w} = \frac{122.7 \ \frac{\text{lbf}}{\text{ft}^3}}{1+0.142} \\ &= 107.4 \text{ lbf/ft}^3 \quad (110 \text{ lbf/ft}^3) \end{aligned}$$

Author Commentary

In geotechnical laboratory testing, mass (recorded in grams) and density (recorded in mass per unit volume) are commonly used. However, in the United States, weight (recorded in pounds-force) and unit weight (recorded in pounds-force per unit volume) are commonly used in soil calculations instead of mass and density.

The answer is (B).

3. Calculate the plasticity index (PI).

$$\text{PI} = \text{LL} - \text{PL} = 25 - 18 = 7$$

Use the USCS chart to determine the major division between coarse- and fine-grained soils using the sieve analysis results. Since 9% of the particles are finer than the no. 200 sieve, 91% are retained on or above the sieve, indicating a coarse-grained soil.

Of the coarse fraction, the percentage of soil retained on the no. 4 sieve is

$$Q_{s,\text{retained}} = (0.41)(0.91) = 0.37 \quad (37\%)$$

Therefore, the percentage of the coarse fraction that passes through the no. 4 sieve is

$$Q_{s,\text{passing}} = 100\% - 37\% = 63\%$$

Over half of the coarse fraction is finer than the no. 4 sieve, so the soil is a sand.

Since the PI is greater than 4, the SM criterion is not met. The percent passing the no. 200 sieve is within the range of 5% to 12%. The USCS chart indicates that this is a borderline case requiring the use of a dual symbol.

Assign the approximate grain diameters for 10% passing, 30% passing, and 60% passing.

D_{10} = no. 200 sieve opening = 0.075 mm

D_{30} = no. 100 sieve opening = 0.150 mm

D_{60} = no. 4 sieve opening = 4.75 mm

Calculate the coefficient of uniformity.

$$C_u = \frac{D_{60}}{D_{10}} = \frac{4.75 \text{ mm}}{0.075 \text{ mm}} = 63$$

Calculate the coefficient of curvature.

$$C_z = \frac{D_{30}^2}{D_{10}D_{60}} = \frac{(0.150 \text{ mm})^2}{(0.075 \text{ mm})(4.75 \text{ mm})} = 0.063$$

The calculated values of C_u and C_z (63 and 0.063, respectively) indicate that the group symbol must start with SP because it does not meet the SW gradation requirements. The Atterberg limits results (LL = 25 and PI = 7) plot on the plasticity chart in the CL-ML zone, which is also above the A-line. Therefore, the second part of the dual symbol is SC.

Author Commentary

When using the USCS chart, be sure to consider all criteria. Certain ranges indicate a dual symbol, but may easily be overlooked.

The answer is (C).

4. Metamorphic rock is generally heavy, dense, and contains rearranged materials. Igneous rock usually contains crystallization without planes and fracturing. The presence of large, flat rock masses of relatively low density and jointing indicates the rock is most likely sedimentary. Furthermore, the presence of well-defined planes and jointing indicates layering, which is more common in sedimentary rock.

Author Commentary

A basic knowledge of rock characteristics such as jointing and fracturing is important in evaluation of slope stability.

The answer is (A).

5. Without soil improvement, the proposed structural loads are too high to be supported by spread footings or a thickened slab-on-grade. These types of foundations could put the building at risk for excessive settlement. A mat and pedestal foundation could be constructed on the claystone, but this option is more costly than drilled cast-in-place concrete piles, since each column location would require excavation.

The relatively shallow depth of the hard claystone permits the installation of drilled cast-in-place concrete piles without excavation. Such piles can be relatively short and permitted to end bear on the claystone layer.

Author Commentary

Knowledge of boring logs and soil profiles is essential for determining what type of foundation is most feasible for a structure.

The answer is (D).

6. Evaluating seepage requires determining the gradation, permeability, and in situ density of the embankment soils. The profile of the dam, including layer positions and thickness, must also be known. While important to the slope stability and structural support of the embankment, soil strength properties such as angle of internal friction, shear strength, and settlement are secondary to the assessment of seepage for this particular investigation. If the dam were to be enlarged, it would be necessary to determine these values.

Author Commentary

Selection of a proper laboratory testing program comes from the basic knowledge of what parameters the tests will generate and what data will be necessary to perform the required evaluation.

The answer is (C).

7. The equation for consolidation settlement is

$$S = C_c\left(\frac{H_o}{1+e_o}\right)\log_{10}\frac{p_o' + \Delta p_v'}{p_o'}$$

Use phase relationships to determine the initial void ratio and the total volume corresponding to the saturated unit weight of the clay. The initial void ratio is

$$e_o = \frac{w(\text{SG})}{S} = \frac{(0.27)(2.69)}{1} = 0.726$$

Find the dry unit weight and the saturated unit weight of the clay.

$$\gamma_{\text{dry}} = \frac{(\text{SG})\gamma_w}{1+e_o} = \frac{(2.69)\left(62.4\ \frac{\text{lbf}}{\text{ft}^3}\right)}{1+0.726} = 97.2\ \text{lbf/ft}^3$$

$$\begin{aligned}\gamma_{\text{sat}} &= \gamma_{\text{dry}}(1+w) = \left(97.2\ \frac{\text{lbf}}{\text{ft}^3}\right)(1+0.27)\\ &= 123.5\ \text{lbf/ft}^3\end{aligned}$$

Calculate the initial effective pressure at the midpoint of the clay layer.

$$\begin{aligned}p_o' &= p_{\text{clay,mid}} - u_o = \gamma_t D_o - \gamma_w D_o = (\gamma_{\text{sat}} - \gamma_w) D_o\\ &= \left(123.5\ \frac{\text{lbf}}{\text{ft}^3} - 62.4\ \frac{\text{lbf}}{\text{ft}^3}\right)(20\ \text{ft})\\ &= 1222\ \text{lbf/ft}^2\end{aligned}$$

Determine the change in effective pressure at the midpoint of the clay layer due to the total increase in surcharge pressure from the fill and the change in water table height.

The surcharge pressure from the fill is

$$p_q = \gamma_{\text{debris}} H_{\text{debris}} = \left(140\ \frac{\text{lbf}}{\text{ft}^3}\right)(2\ \text{ft}) = 280\ \text{lbf/ft}^2$$

The change in effective pressure at the midpoint of the clay layer is

$$\begin{aligned}\Delta p_v' &= p_q + (u_o - u_f) = p_q + (H_{w,o} - H_{w,f})\gamma_w\\ &= 280\ \frac{\text{lbf}}{\text{ft}^2} + (20\ \text{ft} - 4\ \text{ft})\left(62.4\ \frac{\text{lbf}}{\text{ft}^3}\right)\\ &= 1278\ \text{lbf/ft}^2\end{aligned}$$

Calculate the total expected settlement.

$$\begin{aligned}S &= C_c\left(\frac{H_o}{1+e_o}\right)\log_{10}\frac{p_o' + \Delta p_v'}{p_o'}\\ &= (0.29)\left(\frac{40\ \text{ft}}{1+0.726}\right)\log_{10}\frac{1222\ \frac{\text{lbf}}{\text{ft}^2} + 1278\ \frac{\text{lbf}}{\text{ft}^2}}{1222\ \frac{\text{lbf}}{\text{ft}^2}}\\ &\quad \times\left(12\ \frac{\text{in}}{\text{ft}}\right)\\ &= 25\ \text{in}\end{aligned}$$

Author Commentary

To save time, γ_{sat} may be calculated directly instead of calculating γ_{dry} and then adding in the moisture content.

The answer is (C).

8. Determine the mass of water by subtracting the dry mass from the wet mass.

$$m_w = m_{\text{wet}} - m_{\text{dry}} = 37.59\ \text{g} - 24.33\ \text{g} = 13.26\ \text{g}$$

The dry mass of the solids is considered the mass of solids, m_s.

The water content is

$$w = \frac{m_w}{m_s} = \frac{13.26\ \text{g}}{24.33\ \text{g}} = 0.545$$

Calculate the void ratio.

$$e = \frac{w(\text{SG})}{S} = \frac{(0.545)(2.66)}{0.86} = 1.69\quad (1.7)$$

Author Commentary

Be sure to understand the difference between dry mass and wet mass (sometimes referred to as moist mass). The difference between moist and wet is qualitative. Both mean a certain amount of water is present in the soil voids, but the soil is not completely dry nor saturated.

However, terms like dry and saturated can be considered quantitative. For example, dry clearly indicates an absence of moisture in the voids (zero water content), while saturated means the voids are full.

The answer is (C).

9. The dry unit weight, γ_{dry}, and the total volume, V_t, are used to find the weight of the solids in the sample.

$$W_s = \gamma_{\text{dry}} V_t = \left(117.6\ \frac{\text{lbf}}{\text{ft}^3}\right)(1\ \text{ft}^3) = 117.6\ \text{lbf}$$

Calculate the volume of the solids.

$$\frac{W_s}{V_s} = (\text{SG})\gamma_w$$

$$V_s = \frac{W_s}{(\text{SG})\gamma_w} = \frac{117.6\ \text{lbf}}{(2.68)\left(62.4\ \frac{\text{lbf}}{\text{ft}^3}\right)} = 0.70\ \text{ft}^3$$

The volume of voids is

$$V_v = V_t - V_s = 1\ \text{ft}^3 - 0.70\ \text{ft}^3 = 0.30\ \text{ft}^3$$

Since the field sample is saturated, the voids are completely filled with water. The volume of water is equivalent to the volume of voids.

$$V_w = V_v = 0.30\ \text{ft}^3$$

The weight of water in the sample is

$$W_w = \gamma_w V_w = \left(62.4\ \frac{\text{lbf}}{\text{ft}^3}\right)(0.30\ \text{ft}^3) = 18.5\ \text{lbf}$$

Calculate the water content of the sample.

$$w = \frac{W_w}{W_s} \times 100\% = \frac{18.5\ \text{lbf}}{117.6\ \text{lbf}} \times 100\% = 16\%$$

The water content of the soil at the liquid limit is the LL expressed as a percentage.

$$\text{LL} = 38$$
$$w_{\text{LL}} = 38\%$$

Calculate the difference between the water content at the liquid limit and the water content of the field sample on a percent solids basis.

$$w_{\text{diff}} = w_{\text{LL}} - w = 38\% - 16\% = 22\%$$

Author Commentary

A saturated soil is one that has voids perfectly filled with water without bleeding. However, this does not mean there is hydrostatic pore pressure unless a phreatic (water table) symbol is present.

The answer is (B).

10. The soil is at maximum water content when the volume of voids is at 100% saturation. Therefore, the volume of voids is equivalent to the volume of water that the soil can take on without swelling or bleeding.

$$V_w = V_v = V_t - V_s$$

The equation for volume of solids, V_s, is

$$V_s = \frac{W_s}{(\text{SG})\gamma_w}$$

Therefore,

$$V_w = V_t - \frac{W_s}{(\text{SG})\gamma_w} = 0.42\ \text{ft}^3 - \frac{55\ \text{lbf}}{(2.69)\left(62.4\ \frac{\text{lbf}}{\text{ft}^3}\right)} = 0.09\ \text{ft}^3$$

Calculate the corresponding weight of water.

$$W_w = V_w\gamma_w = (0.09\ \text{ft}^3)\left(62.4\ \frac{\text{lbf}}{\text{ft}^3}\right) = 5.8\ \text{lbf}$$

The maximum water content of the soil is

$$w = \frac{W_w}{W_s} \times 100\% = \frac{5.8\ \text{lbf}}{55\ \text{lbf}} \times 100\% = 10.5\% \quad (10\%)$$

Author Commentary

Be sure to understand when a soil can be considered to be under saturated conditions as opposed to buoyant conditions. A soil cannot be assumed to be under buoyant conditions unless it is known that positive pore pressure exists, such as when a water table is present.

The answer is (B).

11. Using a total volume of 1 ft³ for silty sand, determine the volume of solids and the volume of voids using phase relationships. The volume of the solids is

$$e = \frac{V_t}{V_s} - 1$$

$$V_s = \frac{V_t}{e+1} = \frac{1\ \text{ft}^3}{0.73 + 1} = 0.58\ \text{ft}^3$$

The weight of the solids is

$$W_s = V_s\gamma_s = (0.58\ \text{ft}^3)\left(160.3\ \frac{\text{lbf}}{\text{ft}^3}\right) = 93\ \text{lbf}$$

Calculate the volume of the voids.

$$V_v = V_s e = (0.58\ \text{ft}^3)(0.73) = 0.42\ \text{ft}^3$$

Calculate the dry unit weight for the silty sand layer.

$$\gamma_{\text{dry,sand}} = \frac{W_s}{V_t} = \frac{93\ \text{lbf}}{1\ \text{ft}^3} = 93\ \text{lbf/ft}^3$$

The sample is saturated, so the voids are completely filled with water. Therefore, the volume of voids equals the volume of water. Calculate the saturated unit weight for the silty sand layer.

$$\gamma_{\text{sat,sand}} = \frac{W_s + W_w}{V_t} = \frac{W_s + V_v\gamma_w}{V_t}$$
$$= \frac{93 \text{ lbf} + (0.42 \text{ ft}^3)\left(62.4 \frac{\text{lbf}}{\text{ft}^3}\right)}{1 \text{ ft}^3}$$
$$= 119.0 \text{ lbf/ft}^3$$

Calculate the stress for the dry portion of the silty sand (i.e., layer 1). A moisture content is not given for the silty sand. However, since layer 1 is above the water table, capillary action does not exist, and the sand above the water table is dry.

$$\sigma_{\text{dry,sand}} = \gamma_{\text{dry,sand}} z_{\text{layer 1}} = \left(93 \frac{\text{lbf}}{\text{ft}^3}\right)(8 \text{ ft})$$
$$= 741 \text{ lbf/ft}^2$$

Calculate the stress for the saturated portion of the silty sand (i.e., layer 2).

$$\sigma_{\text{sat,sand}} = \gamma_{\text{sat,sand}} z_{\text{layer 2}} = \left(119.0 \frac{\text{lbf}}{\text{ft}^3}\right)(6 \text{ ft})$$
$$= 714 \text{ lbf/ft}^2$$

Calculate the stress at the midpoint of the sandy clay layer (i.e., layer 3).

$$\sigma_{\text{sat,clay}} = \gamma_{\text{sat,clay}} z_{\text{layer 3,mid}} = \left(122.2 \frac{\text{lbf}}{\text{ft}^3}\right)\left(\frac{16 \text{ ft}}{2}\right)$$
$$= 978 \text{ lbf/ft}^2$$

Determine the total stress at the midpoint of the sandy clay layer by finding the sum of the stress values.

$$\sigma_t = \sigma_{\text{dry,sand}} + \sigma_{\text{sat,sand}} + \sigma_{\text{sat,clay}}$$
$$= 741 \frac{\text{lbf}}{\text{ft}^2} + 714 \frac{\text{lbf}}{\text{ft}^2} + 978 \frac{\text{lbf}}{\text{ft}^2}$$
$$= 2433 \text{ lbf/ft}^2$$

The pore pressure at the midpoint of the sandy clay layer is determined using the unit weight of water and the depth of the groundwater based on its location, shown by the triangular phreatic surface symbol.

$$u = \gamma_w z_w = \left(62.4 \frac{\text{lbf}}{\text{ft}^3}\right)\left(6 \text{ ft} + \frac{16 \text{ ft}}{2}\right)$$
$$= 874 \text{ lbf/ft}^2$$

Calculate the effective stress at the midpoint of the sandy clay layer by subtracting the pore pressure from the total stress.

$$\sigma' = \sigma_t - u = 2433 \frac{\text{lbf}}{\text{ft}^2} - 874 \frac{\text{lbf}}{\text{ft}^2}$$
$$= 1559 \text{ lbf/ft}^2 \quad (1600 \text{ lbf/ft}^2)$$

Author Commentary

Incorrect identification of the groundwater level and the associated moisture conditions of the soil layer above the groundwater level can result in miscalculations of the unit weights and stresses for the layers.

The answer is (C).

12. Calculate the cross-sectional area of the standpipe.

$$A_{\text{standpipe}} = \frac{\pi}{4}D^2 = \left(\frac{\pi}{4}\right)(4 \text{ cm})^2 = 12.57 \text{ cm}^2$$

The cross-sectional area of the soil sample is

$$A_{\text{sample}} = wh = (25 \text{ cm})(25 \text{ cm}) = 625 \text{ cm}^2$$

From Darcy's law, the coefficient of permeability is

$$K = 2.3\left(\frac{A_{\text{standpipe}}L}{A_{\text{sample},t}}\right)\log_{10}\frac{h_1}{h_2}$$
$$= (2.3)\left(\frac{(12.57 \text{ cm}^2)(10 \text{ cm})}{(625 \text{ cm}^2)(3 \text{ min})\left(60 \frac{\text{s}}{\text{min}}\right)}\right)$$
$$\times \log_{10}\frac{187 \text{ cm}}{187 \text{ cm} - 51 \text{ cm}}$$
$$= 3.55 \times 10^{-4} \text{ cm/s} \quad (3.6 \times 10^{-4} \text{ cm/s})$$

Author Commentary

Although circular soil specimens are most common, square or rectangular specimens may be used as well. Always check the specimen shape for this type of problem.

The answer is (A).

13. An unconfined compressive strength test requires that the confining pressure be zero. For unconfined conditions, the shear strength is equivalent to the cohesion. The cohesion is half the value of the unconfined compressive strength.

Determine the surface area of the cylindrical concrete sample.

$$A_{\text{concrete}} = \frac{\pi}{4}D^2 = \left(\frac{\pi}{4}\right)(6 \text{ in})^2 = 28.3 \text{ in}^2$$

The unconfined compressive strength is

$$q_u = \frac{P_a}{A_{\text{concrete}}} = \frac{(54 \text{ tons})\left(2000 \ \frac{\text{lbf}}{\text{ton}}\right)}{28.3 \text{ in}^2} = 3820 \text{ lbf/in}^2$$

Calculate the cohesion, which is equivalent to the shear strength for unconfined conditions.

$$c = \frac{q_u}{2} = \frac{3820 \ \frac{\text{lbf}}{\text{in}^2}}{2} = 1910 \text{ lbf/in}^2 \quad (1900 \text{ lbf/in}^2)$$

Author Commentary

Confining pressure is zero when the specimen is tested in axial load only with no radial or cell pressure. This is also known as an unconfined compressive strength test.

The answer is (A).

14. Use Mohr's circle equations to determine the shear stress on the failure plane. The deviator stress is equivalent to the difference between the axial stress and radial stress.

$$\sigma_D = \sigma_A - \sigma_R$$

The angle of the failure plane from the horizontal is doubled for evaluation using Mohr's circle. Calculate the shear stress on the failure plane.

$$\begin{aligned}\tau_\alpha &= \frac{1}{2}(\sigma_A - \sigma_R)\sin 2\alpha = \frac{1}{2}\sigma_D \sin 2\alpha \\ &= \left(\frac{1}{2}\right)\left(2.60 \ \frac{\text{tons}}{\text{ft}^2}\right)\sin\big((2)(61°)\big) \\ &= 1.10 \text{ tons/ft}^2\end{aligned}$$

Author Commentary

Evaluations for shear stresses on failure planes typically require special equations in reference to Mohr's circle. A review of Mohr's circle and related equations is fundamental to solving shear stress problems quickly.

The answer is (B).

15. Calculate the cross-sectional area of the soil sample.

$$\begin{aligned}A &= \frac{\pi}{4}D^2 = \left(\frac{\pi}{4}\right)(45 \text{ cm})^2 \\ &= 1590.4 \text{ cm}^2\end{aligned}$$

Using the density of water, 1 g/cm^3, calculate the volume of water discharged from the sample.

$$\begin{aligned}V &= \frac{m_w}{\rho_w} = \frac{(4 \text{ kg})\left(1000 \ \frac{\text{g}}{\text{kg}}\right)}{1 \ \frac{\text{g}}{\text{cm}^3}} \\ &= 4000 \text{ cm}^3\end{aligned}$$

From Darcy's law, the equation for the coefficient of permeability as determined by the constant-head test procedure is

$$\begin{aligned}K &= \frac{VL}{hAt} = \frac{(4000 \text{ cm}^3)(20 \text{ cm})}{(85 \text{ cm})(1590.4 \text{ cm}^2)(4 \text{ min})\left(60 \ \frac{\text{s}}{\text{min}}\right)} \\ &= 2.47 \times 10^{-3} \text{ cm/s} \quad (2.5 \times 10^{-3} \text{ cm/s})\end{aligned}$$

Author Commentary

Being able to quickly recall basic densities, such as the density of water, can save time during the exam.

The answer is (A).

16. Using phase relationships, determine the wet unit weight and the saturated unit weight of the sand prior to the change in water level.

$$\begin{aligned}\gamma_{\text{wet}} &= \gamma_{\text{dry}}(1 + w) = \left(\frac{W_s}{V_t}\right)(1 + w) \\ &= \left(\frac{108 \text{ lbf}}{1 \text{ ft}^3}\right)(1 + 0.066) \\ &= 115.1 \text{ lbf/ft}^3\end{aligned}$$

Using phase relationships, determine the saturated unit weight of the portion of the sand that becomes submerged after the change in water level. For saturated conditions, the volume of water is equivalent to the volume of voids. From the equation for porosity,

$$n = \frac{V_v}{V_t}$$
$$V_w = V_v = nV_t = (0.28)(1\ \text{ft}^3) = 0.28\ \text{ft}^3$$

Find the weight of the water.

$$W_w = V_w\gamma_w = (0.28\ \text{ft}^3)\left(62.4\ \frac{\text{lbf}}{\text{ft}^3}\right) = 17.5\ \text{lbf}$$

Calculate the saturated unit weight of the submerged sand layer.

$$\gamma_{\text{sat}} = \frac{W_w + W_s}{V_t} = \frac{17.5\ \text{lbf} + 108\ \text{lbf}}{1\ \text{ft}^3} = 125.5\ \text{lbf/ft}^3$$

Prior to the change in water level elevation, the effective stress is equivalent to the total stress at the base of the sand layer, because the pore pressure at the water table is zero.

$$\begin{aligned}\sigma'_{\text{base,prior}} &= \sigma_{\text{wet}} - u_{\text{base}} = \sigma_{\text{wet}} - 0 = \sigma_{\text{wet}} \\ &= \gamma_{\text{wet}}D_{\text{sand}} \\ &= \left(115.1\ \frac{\text{lbf}}{\text{ft}^3}\right)(15\ \text{ft}) \\ &= 1726.9\ \text{lbf/ft}^2\end{aligned}$$

After the change in water level location, the stress of the upper (wet) portion of the sand is added to the stress of the lower (saturated) portion. Since the water table is raised 10 ft, 12 ft of wet sand remain, leaving 3 ft of submerged sand.

$$\begin{aligned}\sigma_{\text{base,after}} &= \sigma_{\text{wet,upper}} + \sigma_{\text{sat,lower}} \\ &= \gamma_{\text{wet}}D_{\text{sand,upper}} + \gamma_{\text{sat}}D_{\text{sand,lower}} \\ &= \left(115.1\ \frac{\text{lbf}}{\text{ft}^3}\right)(12\ \text{ft}) + \left(125.5\ \frac{\text{lbf}}{\text{ft}^3}\right)(3\ \text{ft}) \\ &= 1758.0\ \text{lbf/ft}^2\end{aligned}$$

The effective stress at the base of the sand layer after raising the water level is equivalent to the total stress at the base of the sandy layer less the pore pressure.

$$\begin{aligned}\sigma'_{\text{base,after}} &= \sigma_{\text{base,after}} - u_{\text{base}} = \sigma_{\text{base,after}} - \gamma_w D_w \\ &= 1758.0\ \frac{\text{lbf}}{\text{ft}^2} - \left(62.4\ \frac{\text{lbf}}{\text{ft}^3}\right)(3\ \text{ft}) \\ &= 1570.8\ \text{lbf/ft}^2\end{aligned}$$

Calculate the change in effective stress after raising the water level 10 ft.

$$\begin{aligned}\Delta\sigma'_{\text{base}} &= \sigma'_{\text{base,prior}} - \sigma'_{\text{base,after}} = 1726.9\ \frac{\text{lbf}}{\text{ft}^2} - 1570.8\ \frac{\text{lbf}}{\text{ft}^2} \\ &= 156\ \text{lbf/ft}^2\quad (160\ \text{lbf/ft}^2)\end{aligned}$$

The answer is (A).

17. Since the volumetric flow rate through both soil layers is equal and the flow area through both soil layers is constant,

$$Q_1 = Q_2 = K_1 i_1 A = K_2 i_2 A$$
$$K_1 i_1 = K_2 i_2$$

Use Darcy's law to equate the volumetric flow rates and determine the change in head through each soil layer. The units for the coefficient of permeability cancel out.

$$\begin{aligned}K_1 i_1 &= K_2 i_2 \\ K_1\left(\frac{\Delta h_1}{L_1}\right) &= K_2\left(\frac{\Delta h_2}{L_2}\right) \\ \left(0.6\ \frac{\text{in}}{\text{sec}}\right)\left(\frac{\Delta h_1}{14\ \text{ft}}\right) &= \left(0.4\ \frac{\text{in}}{\text{sec}}\right)\left(\frac{\Delta h_2}{16\ \text{ft}}\right) \\ \Delta h_1 &= (0.58)(\Delta h_2)\end{aligned}$$

When steady-state vertical seepage occurs, hydrostatic conditions do not exist. The sum of the change in head for both layers is equivalent to the total head loss measured at the base of layer 2 and can be used to solve for the change in head through both soil layers.

$$\Delta h_1 + \Delta h_2 = (0.58)(\Delta h_2) + \Delta h_2 = 16\ \text{ft}$$

$$\Delta h_2 = \frac{16\ \text{ft}}{1.58} = 10.11\ \text{ft}$$
$$\Delta h_1 = (0.58)(\Delta h_2) = (0.58)(10.11\ \text{ft}) = 5.89\ \text{ft}$$

Due to steady state seepage, the pore pressure at the base of the layer 1 is equivalent to the piezometric head measured at the base of the layer 1. The piezometric head at the base of the layer 1 can be calculated by subtracting the change in head through layer 1 from the total piezometric head measured at the base of the layer 2.

$$h_1 = L_1 - \Delta h_1 = 14\ \text{ft} - 5.89\ \text{ft} = 8.11\ \text{ft}$$

The pore pressure is

$$u = h_1\gamma_w = (8.11\text{ ft})\left(62.4\ \frac{\text{lbf}}{\text{ft}^3}\right)$$
$$= 506\text{ lbm/ft}^2\quad (510\text{ lbm/ft}^2)$$

Author Commentary

An important aspect of this problem is recognition that the volumetric flow rate is equivalent for both layers. Review groundwater hydrology and seepage for the exam.

The answer is (B).

18. To use the Boussinesq contour chart for both square and circular shapes, the loaded area can be divided into two sections as shown. A_1 is the shaded square, and A_2 is the unshaded half-circle.

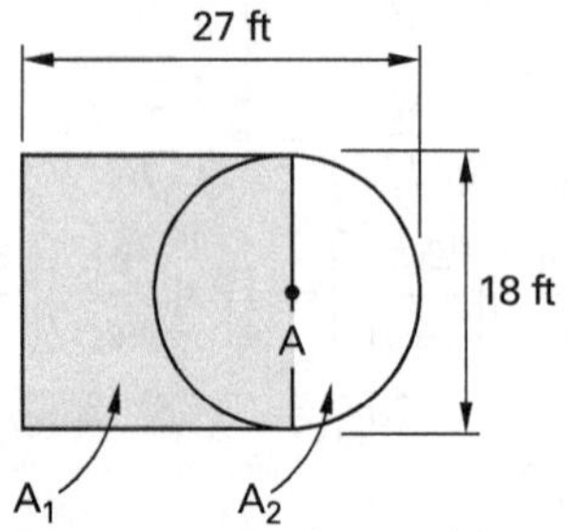

The increase in total vertical pressure (stress) at a depth of 12 ft for each area can be determined using the Boussinesq contour chart. For area A_1, determine the properties required to find the influence factor. The width, B, is 18 ft, and the depth, D, is 12 ft, or $0.7B$.

Point A is located at the edge of area A_1. The influence factor corresponds to the Boussinesq contour located at $0.7B$ down and $0.5B$ laterally from the center of the distributed load depicted by the chart. From the Boussinesq contour chart, the influence factor is 0.35.

$$p_{A_1,12\text{ ft}} = 0.35p_{\text{app}} = (0.35)\left(6500\ \frac{\text{lbf}}{\text{ft}^2}\right) = 2275\text{ lbf/ft}^2$$

The vertical pressure for area A_2 can be determined from the Boussinesq contour chart. However, the shape depicted by the Boussinesq contour chart is a full circle. Therefore, the influence factor for a full circle must be divided in half. The influence factor corresponds to the Boussinesq contour located on the $r/R=0$ line and a depth of $h/R = 12\text{ ft}/9\text{ ft} = 1.3$. From the Boussinesq contour chart, the influence factor is 0.6.

$$p_{A_2,12\text{ ft}} = \frac{0.6p_{\text{app}}}{2} = \frac{(0.6)\left(6500\ \frac{\text{lbf}}{\text{ft}^2}\right)}{2} = 1950\text{ lbf/ft}^2$$

The total increase in vertical pressure at 12 ft below point A is

$$p_{12\text{ ft}} = p_{A_1,12\text{ft}} + p_{A_2,12\text{ft}} = 2275\ \frac{\text{lbf}}{\text{ft}^2} + 1950\ \frac{\text{lbf}}{\text{ft}^2}$$
$$= 4225\text{ lbf/ft}^2\quad (4200\text{ lbf/ft}^2)$$

Author Commentary

The use of superposition, or considering loads one at a time, is a concept that is common to engineering analysis where charts and figures are being used to determine factors. Be familiar with the concept and how to apply it.

The answer is (C).

19. Since the retaining wall has a straight back and granular cohesionless backfill, the Rankine method may be used to find the active earth pressure resultant.

$$k_a = \frac{1-\sin\phi}{1+\sin\phi} = \frac{1-\sin 32^\circ}{1+\sin 32^\circ}$$
$$= 0.31$$

Calculate the active resultant of the triangular pressure distribution.

$$R_a = \frac{1}{2}P_aH = \frac{1}{2}k_a\gamma H^2$$
$$= \left(\frac{1}{2}\right)(0.31)\left(116.4\ \frac{\text{lbf}}{\text{ft}^3}\right)(15\text{ ft})^2$$
$$= 4024\text{ lbf/ft of wall length}$$

The active resultant acts horizontally at $H/3$ above the base of the wall.

The statue weight is considered a point load acting at a distance, x, from the wall. To approximate the value and location of the resultant force acting on the wall, calculate the value of m to use with a surcharge loading chart.

$$m = \frac{x}{H} = \frac{9\text{ ft}}{15\text{ ft}}$$
$$= 0.6$$

For an m value of 0.6, the approximate value of the resultant point load is

$$R_q = \frac{0.46V_q}{H} = \frac{(0.46)(30\text{ tons})\left(2000\ \frac{\text{lbf}}{\text{ton}}\right)}{15\text{ ft}}$$
$$= 1840\text{ lbf/ft of wall length}$$

Referring to a surcharge loading chart, the resultant acts horizontally at $0.4H$ above the base of the wall.

The overturning moment is calculated about the toe using the perpendicular distance from each resultant force to the axis of rotation (lever arms). The calculations are per unit length of wall. Use counterclockwise as the positive direction.

$$M_{\text{toe}} = \frac{H}{3}R_a + 0.4HR_q$$

$$= \frac{\left(\frac{15\text{ ft}}{3}\right)(4024\text{ lbf}) + (0.4)(15\text{ ft})(1840\text{ lbf})}{1000\ \frac{\text{lbf}}{\text{kip}}}$$

$$= 31.2\text{ ft-kips}\quad (31\text{ ft-kips})$$

Author Commentary

Some problems on the exam will require the use of charts common to engineering practice. A list of recommended references is provided in Codes and References. Keep frequently used references available during the exam.

The answer is (C).

20. The Coulomb equation is used for this solution because it allows for sloping backfill, an inclined backfill face, and soil-to-retaining wall friction. Calculate the coefficient of active earth pressure.

$$k_a = \frac{\sin^2(\lambda+\phi)}{\sin^2\lambda\sin(\lambda-\delta)\left(1+\sqrt{\frac{\sin(\phi+\delta)\sin(\phi-\beta)}{\sin(\lambda-\delta)\sin(\lambda+\beta)}}\right)^2}$$

$$= \frac{\sin^2(80° + 27°)}{\sin^2 80°\sin(80° - 22°)\times\left(1+\sqrt{\frac{\sin(27°+22°)\sin(27°-18°)}{\sin(80°-22°)\sin(80°+18°)}}\right)^2}$$

$$= 0.59$$

The active resultant is

$$R_a = \frac{1}{2}k_a\gamma H^2$$

$$= \left(\frac{1}{2}\right)(0.59)\left(117\ \frac{\text{lbf}}{\text{ft}^3}\right)(18\text{ ft})^2$$

$$= 11{,}148\text{ lbf/ft}^2\quad (11{,}000\text{ lbf/ft}^2)$$

Author Commentary

Be certain to know the difference between the Coulomb equation and the Rankine equation, as the application of each depends of the properties of the wall and backfill soil.

The answer is (B).

21. The time factor, T_v, required for 70% consolidation is typically taken as 0.40. Based on the subsurface soil profile, the clay layer is single-drained with a drainage layer thickness of 12 ft. Using the given coefficient of consolidation, the time required for 70% consolidation is

$$T_v = \frac{tC_v}{H_d^2}$$

$$t = \frac{T_vH_d^2}{C_v}$$

$$= \frac{(0.40)(12\text{ ft})^2}{\left(8.9\times10^{-7}\ \frac{\text{ft}^2}{\text{sec}}\right)\left(60\ \frac{\text{sec}}{\text{min}}\right)\left(60\ \frac{\text{min}}{\text{hr}}\right)\left(24\ \frac{\text{hr}}{\text{day}}\right)}$$

$$= 749\text{ days}\quad (750\text{ days})$$

Author Commentary

The index properties given in the soil profile, as well as phase relationship calculations (e.g., for density or unit weight), are not required to solve this problem. Be aware that some problems may give more information than is needed to solve them.

The answer is (C).

22. Calculate the weight of solids by assuming a unit value for the total volume. The weight of solids is determined on a per unit volume basis to find the volume of borrow material required to construct the embankment. Therefore, volume can be omitted from the calculation because it is a unit value, or simply 1.

$$W_t = W_s + W_w$$

$$W_w = wW_s$$

$$W_t = W_s + wW_s$$

$$W_s = \frac{W_t}{1+w} = \frac{102.3\text{ lbf}}{1+0.115} = 91.7\text{ lbf}$$

Use a proportional relationship to calculate the volume of borrow solids needed to construct each cubic foot of embankment. The specifications require 118.8 lbf of solids per cubic foot of embankment fill. Since a unit value is assumed for the volumes, the result of the proportional relationship will be unitless, similar to a fraction.

That result will also reflect the fractional increase of borrow fill volume required to construct a unit volume of the embankment.

$$\frac{W_{s,\text{emb}}}{x_b} = \frac{W_{s,b}}{1}$$

$$x_b = \frac{W_{s,\text{emb}}}{W_{s,b}} = \frac{118.8 \ \frac{\text{lbf}}{\text{ft}^3}}{91.7 \ \frac{\text{lbf}}{\text{ft}^3}}$$
$$= 1.3$$

Use a similar proportional relationship to calculate the volume of borrow materials needed to construct each cubic yard of embankment fill. The assumed value for both the borrow volume and the embankment volume are unit values. However, the proportion of borrow materials required to construct the embankment is a fractional increase as shown.

$$x_b = \frac{V_{b,\text{yd}^3}}{V_{\text{emb,yd}^3}}$$

$$V_{b,\text{yd}^3} = x_b V_{\text{emb,yd}^3} = (1.3)(250{,}000 \ \text{yd}^3)$$
$$= 324{,}000 \ \text{yd}^3 \quad (320{,}000 \ \text{yd}^3)$$

Author Commentary

The use of a unit value for volume saves time and confusion. However, it requires practice to know when assuming a unit value for volume is appropriate.

The answer is (D).

23. Calculate the increase in thickness (vertical volume) of the fill using the swell percentage.

$$\Delta I = t_f S = (2 \ \text{ft})\left(12 \ \frac{\text{in}}{\text{ft}}\right)(0.05) = 1.2 \ \text{in}$$

Author Commentary

Be sure to understand the difference between saturation and inundation. Inundation means that the soil has been flooded with water during laboratory testing. Saturation means that the voids are 100% filled with water. Soil saturation could be determined through laboratory testing or by inspection. Soils below the water table are considered saturated.

The answer is (C).

24. Draw a flow net.

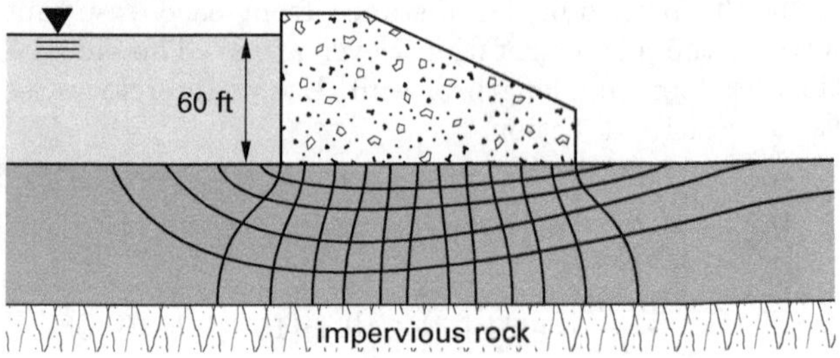

From the flow net, there are 5 flow channels, N_f, and 13 equipotential drops, N_d, between equipotential lines. The values of N_f and N_d are based on knowledge of seepage and associated flow net evaluation. Flow channels represent regions of constant volumetric flow rate, while equipotential drops represent equal differentials in hydraulic energy. In general, the flow channels are parallel to the concrete structure while the equipotential drops are normal to the flow gradient.

From Darcy's law, the seepage rate per foot of dike width is

$$Q = KH\left(\frac{N_f}{N_d}\right) = \left(5 \times 10^{-5} \ \frac{\text{in}}{\text{sec}}\right)\left(\frac{60 \ \text{ft}}{12 \ \frac{\text{in}}{\text{ft}}}\right)\left(\frac{5}{13}\right)$$
$$= 9.6 \times 10^{-5} \ \text{ft}^2/\text{sec}$$

Author Commentary

A knowledge of how to draw flow nets is very important to seepage analysis, and can be useful both in providing a qualitative view and a basis for a quantitative evaluation. Many references present guidelines for drawing flow nets based on Darcy's law and associated graphical representations of fluid mechanics principles.

The answer is (A).

25. The equation for consolidation settlement is

$$S = C_c\left(\frac{H_o}{1+e_o}\right)\log_{10}\frac{p_o' + \Delta p_v'}{p_o'}$$

Calculate the initial effective pressure at the midpoint of the clay layer, layer 3.

$$\begin{aligned}p_o' &= p_{\text{layer 1}} + p_{\text{layer 2}} + p_{\text{layer 3,mid}} - u \\ &= \gamma_{\text{sand,dry}}(1 + w_{\text{layer 1}})D_{\text{layer 1}} \\ &\quad + \gamma_{\text{sand,dry}}(1 + w_{\text{layer 2}})D_{\text{layer 3}} \\ &\quad + \gamma_{\text{clay,dry}}(1 + w_{\text{layer 3}})D_{\text{layer 3,mid}} - \gamma_w D_{w,\text{mid}} \\ &= \left(105.8\ \frac{\text{lbf}}{\text{ft}^3}\right)(1 + 0.074)(11\ \text{ft}) \\ &\quad + \left(105.8\ \frac{\text{lbf}}{\text{ft}^3}\right)(1 + 0.179)(4\ \text{ft}) \\ &\quad + \left(112.5\ \frac{\text{lbf}}{\text{ft}^3}\right)(1 + 0.21)(6.5\ \text{ft}) \\ &\quad - \left(62.4\ \frac{\text{lbf}}{\text{ft}^3}\right)(10.5\ \text{ft}) \\ &= 1978\ \text{lbf/ft}^2\end{aligned}$$

Use the Boussinesq stress contour chart for square foundations to determine the change in stress at the midpoint of the clay layer. Calculate the unit width, B.

$$\left(\frac{11\ \text{ft} + 4\ \text{ft} + 6.5\ \text{ft}}{30\ \text{ft}}\right)B = 0.72B$$

From the Boussinesq stress contour chart, the influence factor corresponding to $0.72B$ is 0.61.

$$\begin{aligned}\Delta p_v' &= 0.61p = 0.61p_{\text{net,center}} = (0.61)\left(2700\ \frac{\text{lbf}}{\text{ft}^2}\right) \\ &= 1647\ \text{lbf/ft}^2\end{aligned}$$

Determine the void ratio of the clay layer.

$$\begin{aligned}e_o &= \frac{w(\text{SG})}{S} = \frac{(0.21)(2.68)}{1} \\ &= 0.56\end{aligned}$$

The expected consolidated settlement due to the placement of the foundation is

$$\begin{aligned}S &= C_c\left(\frac{H_o}{1 + e_o}\right)\log_{10}\frac{p_o' + \Delta p_v'}{p_o'} \\ &= (0.38)\left(\frac{13\ \text{ft}}{1 + 0.56}\right)\log_{10}\left(\frac{1978\ \frac{\text{lbf}}{\text{ft}^2} + 1647\ \frac{\text{lbf}}{\text{ft}^2}}{1978\ \frac{\text{lbf}}{\text{ft}^2}}\right)\left(12\ \frac{\text{in}}{\text{ft}}\right) \\ &= 10\ \text{in}\end{aligned}$$

Author Commentary

Become familiar with the Boussinesq stress contour chart to z problems involving pressure, depth, and consolidation more quickly.

The answer is (D).

26. Rearrange the equation for base shear to calculate the total weight.

$$V = \frac{WS_a}{g}$$

$$\begin{aligned}W &= \frac{Vg}{S_a} = \frac{\left(\frac{900\ \text{lbf}}{1000\ \frac{\text{lbf}}{\text{kip}}}\right)\left(32.2\ \frac{\text{ft}}{\text{sec}^2}\right)}{0.25\ \frac{\text{ft}}{\text{sec}^2}} \\ &= 115.9\ \text{kips}\quad (120\ \text{kips})\end{aligned}$$

The answer is (B).

27. The sign is supported by a cantilever beam consisting of the two metal poles, which are fixed at the foundation and can deflect perpendicular to the plane of the sign. Use the stiffness for a single pole to determine the natural period of vibration. The stiffness required to deflect each pole 1 ft laterally is

$$k = \frac{3EI}{h^3} = \frac{(3)\left(5.9 \times 10^6\ \frac{\text{lbf}}{\text{in}^2}\right)(22{,}000\ \text{in}^4)}{(100\ \text{ft})^3\left(12\ \frac{\text{in}}{\text{ft}}\right)^2} = 2704\ \text{lbf/ft}$$

The two poles are required to move together for the sign to deflect. Therefore, double the stiffness in the calculation of the natural period of vibration.

$$T = 2\pi\sqrt{\frac{W}{gk}} = 2\pi\sqrt{\frac{(20\ \text{kips})\left(1000\ \frac{\text{lbf}}{\text{kip}}\right)}{\left(32.2\ \frac{\text{ft}}{\text{sec}^2}\right)(2)\left(2704\ \frac{\text{lbf}}{\text{ft}}\right)}}$$
$$= 2.1\ \text{sec}$$

Author Commentary

Stiffness is the force required to cause a unit of deflection in a structure. It is used in determining vibration as it applies to the seismic evaluation of a structure. Be sure to review structural analysis when preparing for the geotechnical depth exam.

The answer is (C).

28. Use the tip penetration resistance portion of the cone penetration test (CPT) data to determine the cyclic resistance ratio ($\text{CRR}_{M=7.5}$) after normalization to an equivalent clean sand value and 1 atm of pressure.

Normalization is achieved through iteration using the exponent n, which varies between 0.5 and 1.0 depending on soil type. An n value of 1.0 is appropriate for clayey soil, while an n value of 0.5 is appropriate for clean sand.

Calculate the stress at a depth of 20 ft, effective stress, and CPT soil behavior type index. The stress at a depth of 20 ft is

$$\sigma_{vo} = \gamma_{\text{dry,sand}} D_{\text{dry sand}} + \gamma_{\text{dry,sand}}(1+w)D_{\text{wet sand}}$$
$$= \left(105.3\ \frac{\text{lbf}}{\text{ft}^3}\right)(4\ \text{ft}) + \left(105.3\ \frac{\text{lbf}}{\text{ft}^3}\right)(1+0.14)(16\ \text{ft})$$
$$= 2342\ \text{lbf/ft}^2$$

Find the pore pressure.

$$u = \gamma_w D_{\text{wet sand}} = \left(62.4\ \frac{\text{lbf}}{\text{ft}^3}\right)(16\ \text{ft})$$
$$= 998\ \text{lbf/ft}^2$$

Calculate the effective stress.

$$\sigma'_{vo} = \sigma_{vo} - u = 2342\ \frac{\text{lbf}}{\text{ft}^2} - 998\ \frac{\text{lbf}}{\text{ft}^2}$$
$$= 1343\ \text{lbf/ft}^2$$

Assume an initial value of $n = 1$ and calculate the CPT soil behavior type index. Standard atmospheric pressure is 14.7 lbf/in². The CPT penetration resistance is

$$Q = \left(\left(\frac{q_c - \sigma_{vo}}{p_{\text{atm}}}\right)\left(\frac{p_{\text{atm}}}{\sigma'_{vo}}\right)\right)^n$$
$$= \left(\left(\frac{37{,}000\ \frac{\text{lbf}}{\text{ft}^2} - 2342\ \frac{\text{lbf}}{\text{ft}^2}}{\left(14.7\ \frac{\text{lbf}}{\text{in}^2}\right)\left(12\ \frac{\text{in}}{\text{ft}}\right)^2}\right)\left(\frac{\left(14.7\ \frac{\text{lbf}}{\text{in}^2}\right)\left(12\ \frac{\text{in}}{\text{ft}}\right)^2}{1343\ \frac{\text{lbf}}{\text{ft}^2}}\right)\right)^1$$
$$= 25.8$$

The normalized friction ratio is

$$F = \frac{f_s}{q_c - \sigma_{vo}} \times 100\% = \frac{f_R q_c}{q_c - \sigma_{vo}} \times 100\%$$
$$= \frac{(0.012)\left(37{,}000\ \frac{\text{lbf}}{\text{ft}^2}\right)}{37{,}000\ \frac{\text{lbf}}{\text{ft}^2} - 2342\ \frac{\text{lbf}}{\text{ft}^2}} \times 100\%$$
$$= 1.28\%$$

The CPT soil behavior type index is

$$I_c = \sqrt{(3.47 - \log_{10} Q)^2 + (1.22 + \log_{10} F)^2}$$
$$= \sqrt{(3.47 - \log_{10} 25.8)^2 + (1.22 + \log_{10} 1.28)^2}$$
$$= 2.4$$

The CPT soil behavior type index using $n = 1$ is less than 2.6. Recalculate using $n = 0.5$.

$$Q = \left(\left(\frac{q_c - \sigma_{vo}}{p_{\text{atm}}}\right)\left(\frac{p_{\text{atm}}}{\sigma'_{vo}}\right)\right)^n$$
$$= \left(\left(\frac{37{,}000\ \frac{\text{lbf}}{\text{ft}^2} - 2342\ \frac{\text{lbf}}{\text{ft}^2}}{\left(14.7\ \frac{\text{lbf}}{\text{in}^2}\right)\left(12\ \frac{\text{in}}{\text{ft}}\right)^2}\right)\left(\frac{\left(14.7\ \frac{\text{lbf}}{\text{in}^2}\right)\left(12\ \frac{\text{in}}{\text{ft}}\right)^2}{1343\ \frac{\text{lbf}}{\text{ft}^2}}\right)\right)^{0.5}$$
$$= 20.6$$

The normalized friction ratio remains unchanged since it is not dependent on the n value. Calculate the CPT soil behavior type index for $n = 0.5$.

$$I_c = \sqrt{(3.47 - \log_{10} Q)^2 + (1.22 + \log_{10} F)^2}$$
$$= \sqrt{(3.47 - \log_{10} 20.6)^2 + (1.22 + \log_{10} 1.28)^2}$$
$$= 2.3$$

The CPT soil behavior type index remains less than 2.6, so it can be used to determine the correction for grain characteristics, K_c. The equation for K_c depends on the value of I_c.

$$K_c = 1.0 \quad [\text{for } I_c \leq 1.64]$$

$$K_c = -0.403I_c^4 + 5.58I_c^3 - 21.63I_c^2 + 33.75I_c - 17.88 \quad [\text{for } I_c > 1.64]$$

The CPT soil behavior type index is greater than 1.64. Therefore,

$$\begin{aligned} K_c &= -0.403I_c^4 + 5.58I_c^3 - 21.63I_c^2 + 33.75I_c - 17.88 \\ &= (-0.403)(2.3)^4 + (5.58)(2.3)^3 - (21.63)(2.3)^2 \\ &\quad + (33.75)(2.3) - 17.88 \\ &= 1.81 \end{aligned}$$

Using $n = 0.5$, calculate the factor C_Q to normalize the tip penetration resistance value to 1 atm of pressure.

$$\begin{aligned} C_Q &= \left(\frac{p_{\text{atm}}}{\sigma'_{vo}}\right)^n \leq 1.7 = \left(\frac{\left(14.7\ \frac{\text{lbf}}{\text{in}^2}\right)\left(12\ \frac{\text{in}}{\text{ft}}\right)^2}{1343\ \frac{\text{lbf}}{\text{ft}^2}}\right)^{0.5} \\ &= 1.3 \quad (\leq 1.7) \end{aligned}$$

The tip penetration resistance is

$$\begin{aligned} q_{c1N} &= C_Q\left(\frac{q_c}{p_{\text{atm}}}\right) = (1.3)\left(\frac{37{,}000\ \frac{\text{lbf}}{\text{ft}^2}}{\left(14.7\ \frac{\text{lbf}}{\text{in}^2}\right)\left(12\ \frac{\text{in}}{\text{ft}}\right)^2}\right) \\ &= 21.9 \end{aligned}$$

Normalized for clean sand,

$$(q_{c1N})_{cs} = K_c q_{c1N} = (1.81)(21.9) = 40$$

Determine the applicable equation for calculating $\text{CRR}_{M=7.5}$ based on plotted curve data found in standard reference tables.

$$\text{CRR}_{M=7.5} = (0.833)\left(\frac{(q_{c1N})_{cs}}{1000}\right) + 0.05 \quad [\text{for } (q_{c1N})_{cs} < 50]$$

$$\text{CRR}_{M=7.5} = (93)\left(\frac{(q_{c1N})_{cs}}{1000}\right)^3 + 0.08 \quad [\text{for } 50 \leq (q_{c1N})_{cs} < 160]$$

Since $(q_{c1N})_{cs}$ is equal to 40, which is less than 50, the value for $\text{CRR}_{M=7.5}$ is

$$\begin{aligned} \text{CRR}_{M=7.5} &= (0.833)\left(\frac{(q_{c1N})_{cs}}{1000}\right) + 0.05 \\ &= (0.833)\left(\frac{40}{1000}\right) + 0.05 \\ &= 0.083 \quad (0.08) \end{aligned}$$

The answer is (C).

29. Find Marston's constant for broad fill, C_P, using the ratio of the depth of the embankment to the trench width.

$$\frac{h}{D} = \frac{6\ \text{ft}}{\frac{18\ \text{in}}{12\ \frac{\text{in}}{\text{ft}}}} = 4$$

For flexible pipe and average conditions with an h/D value of 4, $C_P = 5.4$. The dead load on the buried pipe is

$$\begin{aligned} w_d &= C_P\gamma D^2 = (5.4)\left(115\ \frac{\text{lbf}}{\text{ft}^3}\right)\left(\frac{18\ \text{in}}{12\ \frac{\text{in}}{\text{ft}}}\right)^2 \\ &= 1397\ \text{lbf/ft} \quad (1400\ \text{lbf/ft}) \end{aligned}$$

Author Commentary

Be sure to use Marston's constant for flexible pipe, not rigid pipe.

The answer is (D).

30. Use the Taylor slope stability chart for an angle of internal friction of 0° to determine the factor of safety with respect to cohesion. From the Taylor slope stability chart, the depth factor, d, is 0.5, and the stability number, N_o, is 6.

Calculate the corresponding cohesive factor of safety. Since the slope is submerged, the effective unit weight (found from the buoyant unit weight) must be used when calculating the factor of safety.

$$F_{\text{cohesive}} = \frac{N_o c}{\gamma_{\text{eff}} H} = \frac{N_o c}{(\gamma_{\text{dry}}(1+w) - \gamma_w)H}$$

$$= \frac{(6)\left(230\ \frac{\text{lbf}}{\text{ft}^2}\right)}{\left(\left(99\ \frac{\text{lbf}}{\text{ft}^3}\right)(1+0.12) - 62.4\ \frac{\text{lbf}}{\text{ft}^3}\right)(20\ \text{ft})}$$

$$= 1.42 \quad (1.4)$$

Author Commentary

Since the slope is submerged, the soil has positive pore pressure. Therefore, calculating the effective unit weight from the buoyant unit weight is critical to the solution.

The answer is (C).

31. Find the Terzaghi bearing capacity factors N_γ and N_q using standard reference tables. The bearing capacity factor for cohesion, N_c, is not required since a value for cohesion is not given. For $\phi = 34°$, the applicable bearing capacity factors are

$$N_\gamma = 35.0$$
$$N_q = 36.5$$

Calculate the ultimate bearing capacity. The additional surface surcharge, p_q, is not required for this problem, so use a value of zero. Disregard cohesion.

$$q_{\text{ult}} = \frac{1}{2}\gamma B N_\gamma + cN_c + (p_q + \gamma D_f)N_q$$

$$= \left(\frac{1}{2}\right)\left(111.1\ \frac{\text{lbf}}{\text{ft}^3}\right)(3\ \text{ft})(35.0) + \left(0\ \frac{\text{lbf}}{\text{ft}^2}\right)(0) + \left(0\ \frac{\text{lbf}}{\text{ft}^3} + \left(111.1\ \frac{\text{lbf}}{\text{ft}^3}\right)(1.5\ \text{ft})\right)(36.5)$$

$$= 11{,}916\ \text{lbf/ft}^2$$

The net bearing capacity is

$$q_{\text{net}} = q_{\text{ult}} - \gamma D_f = 11{,}916\ \frac{\text{lbf}}{\text{ft}^2} - \left(111.1\ \frac{\text{lbf}}{\text{ft}^3}\right)(1.5\ \text{ft})$$

$$= 11{,}749\ \text{lbf/ft}^2 \quad (12{,}000\ \text{lbf/ft}^2)$$

Author Commentary

This problem provides a good test of engineering judgment through evaluating the capacity of a system. It is important to recognize that the shallower embedment side will fail first.

The answer is (B).

32. Draw a flow net.

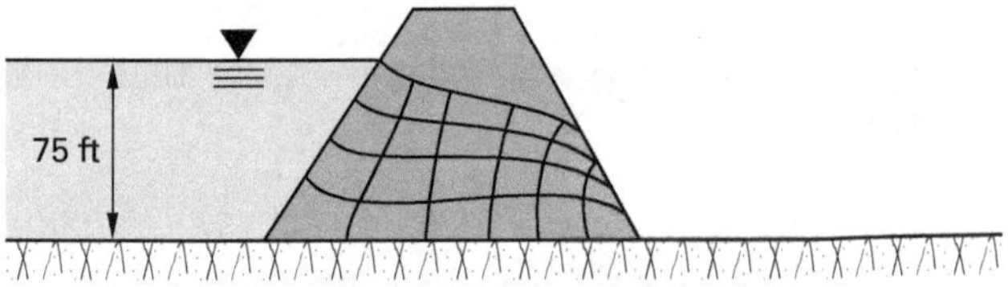

From the flow net, there are 4 flow channels, N_f, and 6 equipotential drops, N_d, between equipotential lines. The values of N_f and N_d are based on flow net evaluation. Flow channels represent regions of constant volumetric flow rate, and equipotential drops represent equal divisions in hydraulic energy. In general, the flow channels parallel the concrete structure while the equipotential drops are normal to the flow gradient.

From Darcy's law, the seepage rate is

$$Q = KH\left(\frac{N_f}{N_d}\right) = \frac{\left(6 \times 10^{-7}\ \frac{\text{in}}{\text{sec}}\right)(75\ \text{ft})\left(\frac{4}{6}\right)}{12\ \frac{\text{in}}{\text{ft}}}$$

$$= 2.5 \times 10^{-6}\ \text{ft}^2/\text{sec}$$

Author Commentary

The accuracy of a flow net depends greatly on the shape of the seepage area. One of the best methods for drawing a flow net is by referencing an example of a similar shape.

The answer is (C).

33. Based on the pipe dead load strength (crushing strength), bedding class, and factor of safety, calculate the total load allowable per foot of pipe length. From standard reference tables, the bedding load factor, LF, for class C bedding is 1.5.

$$w_{\text{allowable}} = (\text{pipe crushing strength})\left(\frac{\text{LF}}{F}\right)$$

$$= ((\text{D-load strength})D_i)\left(\frac{\text{LF}}{F}\right)$$

$$= \frac{\left(750\ \frac{\text{lbf}}{\text{ft}}\right)(30\ \text{in})\left(\frac{1.5}{1.5}\right)}{12\ \frac{\text{in}}{\text{ft}}}$$

$$= 1875\ \text{lbf/ft}$$

Find Marston's constant (buried pipe loading coefficient), C, using the ratio of depth of backfill (measured from the top of the pipe) to trench width and the type of backfill material.

$$\frac{h}{B} = \frac{4\ \text{ft}}{4\ \text{ft}} = 1$$

For an h/B value of 1 and sand backfill, $C = 0.84$. Use Marston's formula to calculate the actual dead load on the buried pipe.

$$w_d = C\gamma B^2 = (0.84)\left(100\ \frac{\text{lbf}}{\text{ft}^3}\right)(4\ \text{ft})^2 = 1344\ \text{lbf/ft}$$

The maximum live load that can be applied at the top of the pipe is the remaining capacity available after subtracting the dead load from the total allowable load.

$$\begin{aligned} w_l &= w_{\text{allowable}} - w_d = 1875\ \frac{\text{lbf}}{\text{ft}} - 1344\ \frac{\text{lbf}}{\text{ft}} \\ &= 531\ \text{lbf/ft} \quad (530\ \text{lbf/ft}) \end{aligned}$$

Author Commentary

Tab applicable factors and values, such as Marston's constants, in your reference material for easy access during the exam.

The answer is (A).

34. Use the SPT data to calculate the cyclic resistance ratio ($\text{CRR}_{M=7.5}$).

$$\begin{aligned} \text{CRR}_{M=7.5} &= \frac{1}{34-(N_1)_{60,\text{cs}}} + \frac{(N_1)_{60,\text{cs}}}{135} \\ &\quad + \frac{50}{(10(N_1)_{60,\text{cs}}+45)^2} - \frac{1}{200} \end{aligned}$$

The SPT data for the depth of 22 ft should be corrected for effective overburden pressure and fines content. The correction requires a comparison of the initial effective stress to atmospheric pressure, 2000 lbf/ft^2.

$$\begin{aligned} N_{60} &= N_{60,\text{2nd 6 in}} + N_{60,\text{3rd 6 in}} = 11 + 11 \\ &= 22 \end{aligned}$$

Calculate the initial effective stress.

$$\begin{aligned} \sigma'_{vo} &= \gamma_{\text{dry,sand}} D_{\text{dry sand}} + \gamma_{\text{dry,sand}}(1+w)(D_{\text{wet sand,23 ft}}) \\ &\quad - \gamma_w D_{w,23\ \text{ft}} \\ &= \left(109.4\ \frac{\text{lbf}}{\text{ft}^3}\right)(7\ \text{ft}) + \left(109.4\ \frac{\text{lbf}}{\text{ft}^3}\right)(1+0.10)(16\ \text{ft}) \\ &\quad - \left(62.4\ \frac{\text{lbf}}{\text{ft}^3}\right)(16\ \text{ft}) \\ &= 1693\ \text{lbf/ft}^2 \end{aligned}$$

Use the calculated initial effective stress and the atmospheric pressure to find $(N_1)_{60}$.

$$\begin{aligned} (N_1)_{60} &= N_{60}\left(\frac{2.2}{1.2 + \frac{\sigma'_{vo}}{p_{\text{atm}}}}\right) \\ &= 22\left(\frac{2.2}{1.2 + \frac{1693\ \frac{\text{lbf}}{\text{ft}^2}}{2000\ \frac{\text{lbf}}{\text{ft}^2}}}\right) \\ &= 24 \end{aligned}$$

Since the fines content is less than 35% (12% passing),

$$\begin{aligned} \alpha &= e^{1.76 - \frac{190}{\text{fines}^2}} \\ \beta &= 0.99 + \frac{\text{fines}^{1.5}}{1000} \end{aligned}$$

Therefore, normalizing the SPT data for clean sand,

$$\begin{aligned} (N_1)_{60,\text{cs}} &= \alpha + \beta(N_1)_{60} \\ &= e^{1.76 - \frac{190}{\text{fines}^2}} + \left(0.99 + \frac{\text{fines}^{1.5}}{1000}\right)(N_1)_{60} \\ &= e^{1.76 - \frac{190}{(12)^2}} + \left(0.99 + \frac{(12)^{1.5}}{1000}\right)(24) \\ &= 26 \end{aligned}$$

Determine the $CRR_{M=7.5}$.

$$\begin{aligned} CRR_{M=7.5} &= \frac{1}{34-(N_1)_{60,cs}} + \frac{(N_1)_{60,cs}}{135} \\ &\quad + \frac{50}{(10(N_1)_{60,cs}+45)^2} - \frac{1}{200} \\ &= \frac{1}{34-26} + \frac{26}{135} + \frac{50}{((10)(26)+45)^2} - \frac{1}{200} \\ &= 0.30 \end{aligned}$$

Author Commentary

The normalization of SPT results to clean sand is typically required prior to their use in calculations.

The answer is (C).

35. Use Terzaghi bearing capacity tables to determine the factors N_γ, N_c, and N_q. For $\phi = 20°$, the Terzaghi bearing capacity factors are

$$\begin{aligned} N_\gamma &= 5.0 \\ N_c &= 17.7 \\ N_q &= 7.4 \end{aligned}$$

The shape factors, S_γ and S_c, for circular footings are

$$\begin{aligned} S_\gamma &= 0.70 \\ S_c &= 1.20 \end{aligned}$$

Using the Terzaghi-Meyerhof equation, the Terzaghi bearing capacity and shape factors, and the additional surface surcharge, p_q, calculate the bearing capacity.

$$\begin{aligned} q_{ult} &= \frac{1}{2}\gamma B N_\gamma S_\gamma + c N_c S_c + (p_q + \gamma D_f) N_q \\ &= \left(\frac{1}{2}\right)\left(109.4\ \frac{\text{lbf}}{\text{ft}^3}\right)(5\ \text{ft})(5.0)(0.70) \\ &\quad + \left(300\ \frac{\text{lbf}}{\text{ft}^2}\right)(17.7)(1.20) \\ &\quad + \left(\frac{\left(145\ \frac{\text{lbf}}{\text{ft}^3}\right)(4\ \text{in})}{12\ \frac{\text{in}}{\text{ft}}} + \left(109.4\ \frac{\text{lbf}}{\text{ft}^3}\right)(2\ \text{ft})\right)(7.4) \\ &= 9306\ \text{lbf/ft}^2 \end{aligned}$$

Calculate the actual soil contact pressure under the footing due to the applied column load.

$$\begin{aligned} p_{actual} &= \frac{P_{col}}{A_f} = \frac{P_{col}}{\frac{1}{4}\pi B^2} = \frac{(120\ \text{kips})\left(1000\ \frac{\text{lbf}}{\text{kip}}\right)}{\frac{1}{4}\pi(5\ \text{ft})^2} \\ &= 6112\ \text{lbf/ft}^2 \end{aligned}$$

The factor of safety of the column loading with respect to the ultimate bearing capacity is

$$F = \frac{q_{ult}}{p_{actual}} = \frac{9306\ \frac{\text{lbf}}{\text{ft}^2}}{6112\ \frac{\text{lbf}}{\text{ft}^2}} = 1.5$$

Author Commentary

Be sure to understand the factors required to solve each problem. Recognize when they are given in the problem statement and when they must be found in reference materials.

The answer is (A).

36. Use Meyerhof bearing capacity tables to determine the factors N_γ, N_c, and N_q. For $\phi = 30°$, the Meyerhof bearing capacity factors are

$$\begin{aligned} N_\gamma &= 15.7 \\ N_c &= 30.1 \\ N_q &= 18.4 \end{aligned}$$

The shape factors, S_γ and S_c, for rectangular footings are

$$\begin{aligned} S_\gamma &= 0.90 \\ S_c &= 1.12 \end{aligned}$$

Calculate the ultimate bearing capacity with adjustment for buoyancy. The value of p_q is zero since there is no additional surface surcharge.

$$q_{ult} = \frac{1}{2}\gamma_b B N_\gamma S_\gamma + c N_c S_c + \left(p_q + \gamma_{dry} D_f + \gamma_w (D_w - D_f)\right) N_q$$

$$= \left(\frac{1}{2}\right)\left(\left(102.3\ \frac{\text{lbf}}{\text{ft}^3}\right)(1 + 0.13) - 62.4\ \frac{\text{lbf}}{\text{ft}^3}\right) \times (4\ \text{ft})(15.7)(0.90) + \left(150\ \frac{\text{lbf}}{\text{ft}^2}\right)(30.1)(1.12)$$

$$+ \left(0\ \frac{\text{lbf}}{\text{ft}^3} + \left(102.3\ \frac{\text{lbf}}{\text{ft}^3}\right)(3.5\ \text{ft}) + \left(62.4\ \frac{\text{lbf}}{\text{ft}^3}\right)(2\ \text{ft} - 3.5\ \text{ft})\right)(18.4)$$

$$= 11{,}426\ \text{lbf/ft}^2 \quad (11{,}400\ \text{lbf/ft}^2)$$

The answer is (C).

37. Determine the factor of safety with respect to cohesion using the Taylor slope stability chart for an angle of internal friction of 0°. Since a depth, D, to a firm base beneath the slope is not given, assume the depth is infinite. Therefore, the depth factor, d, is also infinite. From the Taylor slope stability chart, the stability number, N_o, is 5.53.

Calculate the corresponding cohesive factor of safety.

$$F_{cohesive} = \frac{N_o c}{\gamma H} = \frac{(5.53)\left(450\ \frac{\text{lbf}}{\text{ft}^2}\right)}{\left(119\ \frac{\text{lbf}}{\text{ft}^3}\right)(30\ \text{ft})} = 0.70$$

Author Commentary

Be certain to recognize when more specific information is required and when it isn't required. Knowing the type of unit weight (e.g., wet, saturated, or dry) is not necessary for solving this problem because neither the moisture conditions nor a water table level is given.

The answer is (A).

38. The mass of water is calculated by subtracting the dry mass plus container value, $m_{dry+cont}$, from the wet mass plus container value, $m_{wet+cont}$.

$$m_w = m_{wet+cont} - m_{dry+cont} = 27.2\ \text{g} - 21.8\ \text{g} = 5.4\ \text{g}$$

The dry mass of the sample is determined by subtracting the mass of the container, m_{cont}, from the dry mass plus container value.

$$m_{s,dry} = m_{dry+cont} - m_{cont} = 21.8\ \text{g} - 1.34\ \text{g} = 20.5\ \text{g}$$

The water content of the soil is determined as a percentage and is calculated as the mass of water contained in the sample divided by the sample's dry mass.

$$w = \frac{m_w}{m_{dry}} \times 100\% = \frac{5.4\ \text{g}}{20.5\ \text{g}} \times 100\% = 26.4\% \quad (26\%)$$

Author Commentary

Phase relationship evaluations typically begin with an understanding of basic properties such as water content. Knowing the basic laboratory determination of water content is fundamental to geotechnical practice problems.

The answer is (D).

39. The calculation for Richter magnitude uses the active fault length.

$$\log_{10} L = 1.02M - 5.77$$

$$M = \frac{\log_{10} L + 5.77}{1.02} = \frac{\log_{10}(6\ \text{km}) + 5.77}{1.02} = 6.4$$

The answer is (C).

40. Use the equation for consolidation settlement to determine the settlement that results from construction of the raft foundation.

$$S = C_c\left(\frac{H_o}{1 + e_o}\right)\log_{10}\frac{p'_{vo} + \Delta p'_v}{p'_{vo}}$$

Determine the initial effective pressure at the midpoint of the clay layer.

$$p'_{vo} = p_{clay,dry} + p_{clay,mid,wet} - u$$

$$= \gamma_{dry} D_{clay} + \gamma_{dry}(1 + w) D_{clay,mid} - \gamma_w D_{mid}$$

$$= \left(111.1\ \frac{\text{lbf}}{\text{ft}^3}\right)(10\ \text{ft}) + \left(111.1\ \frac{\text{lbf}}{\text{ft}^3}\right)(1 + 0.173)(5\ \text{ft}) - \left(62.4\ \frac{\text{lbf}}{\text{ft}^3}\right)(5\ \text{ft})$$

$$= 1451\ \text{lbf/ft}^2$$

To determine the change in effective pressure, subtract the total pressure due to the soil excavation from the additional pressure of the new structural load. The self-weight of the below-grade walls and raft foundation elements must be included in the total structural load.

$$\begin{aligned}
\Delta p_v' &= q_{\text{total load}} - p_{\text{clay,dry,9 ft}} \\
&= q_{\text{superstructure load}} + q_{\text{walls+raft}} - \sigma_{\text{clay,dry,9 ft}} \\
&= q_{\text{superstructure load}} \\
&\quad + \frac{\gamma_{\text{conc}}\left(A_{\text{raft}} - (L_{\text{wall}} - 2t_{\text{wall}})^2\right)H_{\text{wall}}}{A_{\text{raft}}} \\
&\quad + \gamma_{\text{conc}} t_{\text{raft}} - \gamma_{\text{dry}} D_{\text{bearing}} \\
&= 1310\ \frac{\text{lbf}}{\text{ft}^2} \\
&\quad + \frac{\left(150\ \frac{\text{lbf}}{\text{ft}^3}\right)\left((120\ \text{ft})^2 - (118\ \text{ft})^2\right)(8\ \text{ft})}{(120\ \text{ft})^2} \\
&\quad + \left(150\ \frac{\text{lbf}}{\text{ft}^3}\right)(1\ \text{ft}) - \left(111.1\ \frac{\text{lbf}}{\text{ft}^3}\right)(9\ \text{ft}) \\
&= 499.8\ \text{lbf/ft}^2
\end{aligned}$$

Determine the total expected consolidation settlement due to the raft foundation and structural load.

$$\begin{aligned}
S &= C_c\left(\frac{H_o}{1+e_o}\right)\log_{10}\frac{p_{vo}' + \Delta p_v'}{p_{vo}'} \\
&= (0.550)\left(\frac{10\ \text{ft}}{1+0.250}\right)\log_{10}\frac{1451\ \frac{\text{lbf}}{\text{ft}^2} + 499.8\ \frac{\text{lbf}}{\text{ft}^2}}{1451\ \frac{\text{lbf}}{\text{ft}^2}} \\
&\quad \times\left(12\ \frac{\text{in}}{\text{ft}}\right) \\
&= 6.8\ \text{in} \quad (7\ \text{in})
\end{aligned}$$

The answer is (C).

Solutions

Practice Exam 2

41. The footings are a common size, with a width greater than 2–4 ft, and groundwater is not present within the zone of influence. Therefore, the SPT result can be used to determine the soil's allowable bearing pressure.

The soil unit weight is close to 100 lbf/ft^3, the SPT correction factor. Therefore, from an SPT correction factor table, C_n can be taken as 1. The allowable soil pressure is

$$\begin{aligned} q_a &= 0.11 C_n N \\ &= (0.11)(1)(12) \\ &= 1.3 \text{ tons/ft}^2 \end{aligned}$$

Author Commentary

Tab your reference material to find SPT correction factors quickly during the exam.

The answer is (D).

42. The total pile capacity (ultimate static bearing capacity) is the sum of the point-bearing capacity (tip capacity) and the skin-friction capacity (side capacity).

$$Q_{\text{ult}} = Q_p + Q_f$$

Use the β-method for cohesive clay, and calculate the skin-friction factor, f_s, and skin-friction capacity, Q_f.

$$\begin{aligned} f_s &= \beta \sigma'_{\text{ave}} \\ Q_f &= f_s \pi B L \end{aligned}$$

For each soil layer, the skin-friction capacity is estimated as a fraction of the average effective vertical stress, which is taken as the value midway through that layer. Values of β are correlated with pile length. Calculate the average effective vertical stress for each layer.

For layer 1,

$$\begin{aligned} \sigma'_{1,\text{ave}} &= \sigma'_{1,\text{mid}} = \sigma_{1,\text{mid}} - u_{1,\text{mid}} \\ &= \gamma_{\text{dry},1}\left(\frac{L_1}{2}\right) - 0 \\ &= \left(107.5 \ \frac{\text{lbf}}{\text{ft}^3}\right)\left(\frac{10 \text{ ft}}{2}\right) - 0 \\ &= 537.5 \text{ lbf/ft}^2 \end{aligned}$$

For layer 2,

$$\begin{aligned} \sigma'_{2,\text{ave}} &= \gamma_{\text{dry},1} L_1 + (\gamma_{\text{dry},1}(1 + w_2) - \gamma_w)\left(\frac{L_2}{2}\right) \\ &= \left(107.5 \ \frac{\text{lbf}}{\text{ft}^3}\right)(10 \text{ ft}) \\ &\quad + \left(\begin{array}{c} \left(107.5 \ \frac{\text{lbf}}{\text{ft}^3}\right)(1 + 0.198) \\ -62.4 \ \frac{\text{lbf}}{\text{ft}^3} \end{array}\right)\left(\frac{4 \text{ ft}}{2}\right) \\ &= 1207.8 \text{ lbf/ft}^2 \end{aligned}$$

For layer 3,

$$\begin{aligned} \sigma'_{3,\text{ave}} &= \gamma_{\text{dry},1} L_1 + (\gamma_{\text{dry},1}(1 + w_2) - \gamma_w) L_2 \\ &\quad + (\gamma_{\text{sat},3} - \gamma_w)\left(\frac{L_3}{2}\right) \\ &= \left(107.5 \ \frac{\text{lbf}}{\text{ft}^3}\right)(10 \text{ ft}) \\ &\quad + \left(\left(107.5 \ \frac{\text{lbf}}{\text{ft}^3}\right)(1 + 0.198) - 62.4 \ \frac{\text{lbf}}{\text{ft}^3}\right)(4 \text{ ft}) \\ &\quad + \left(111.8 \ \frac{\text{lbf}}{\text{ft}^3} - 62.4 \ \frac{\text{lbf}}{\text{ft}^3}\right)\left(\frac{16 \text{ ft}}{2}\right) \\ &= 1735.8 \text{ lbf/ft}^2 \end{aligned}$$

The pile extends only 10 ft into layer 4, so use 10 ft when calculating the midpoint of layer 4.

$$\begin{aligned}\sigma'_{4,\text{ave}} &= \gamma_{\text{dry},1}L_1 + (\gamma_{\text{dry},1}(1+w_2) - \gamma_w)L_2 \\ &\quad + (\gamma_{\text{sat},3} - \gamma_w)L_3 + (\gamma_{\text{sat},4} - \gamma_w)\left(\frac{L_4}{2}\right) \\ &= \left(107.5\ \frac{\text{lbf}}{\text{ft}^3}\right)(10\ \text{ft}) \\ &\quad + \left(\left(107.5\ \frac{\text{lbf}}{\text{ft}^3}\right)(1+0.198) - 62.4\ \frac{\text{lbf}}{\text{ft}^3}\right)(4\ \text{ft}) \\ &\quad + \left(111.8\ \frac{\text{lbf}}{\text{ft}^3} - 62.4\ \frac{\text{lbf}}{\text{ft}^3}\right)(16\ \text{ft}) \\ &\quad + \left(113.3\ \frac{\text{lbf}}{\text{ft}^3} - 62.4\ \frac{\text{lbf}}{\text{ft}^3}\right)\left(\frac{10\ \text{ft}}{2}\right) \\ &= 2385.4\ \text{lbf/ft}^2\end{aligned}$$

For each layer, the length of pile is less than 50 ft. Therefore, the β value for each layer is 0.3.

Calculate the skin-friction factor for each layer.

For layer 1,

$$\begin{aligned}f_{s,1} &= \beta\sigma'_{1,\text{ave}} = (0.3)\left(537.5\ \frac{\text{lbf}}{\text{ft}^2}\right) \\ &= 161.3\ \text{lbf/ft}^2\end{aligned}$$

For layer 2,

$$\begin{aligned}f_{s,2} &= \beta\sigma'_{2,\text{ave}} = (0.3)\left(1207.8\ \frac{\text{lbf}}{\text{ft}^2}\right) \\ &= 362.3\ \text{lbf/ft}^2\end{aligned}$$

For layer 3,

$$\begin{aligned}f_{s,3} &= \beta\sigma'_{3,\text{ave}} = (0.3)\left(1735.8\ \frac{\text{lbf}}{\text{ft}^2}\right) \\ &= 520.7\ \text{lbf/ft}^2\end{aligned}$$

For layer 4,

$$\begin{aligned}f_{s,4} &= \beta\sigma'_{4,\text{ave}} = (0.3)\left(2385.4\ \frac{\text{lbf}}{\text{ft}^2}\right) \\ &= 715.6\ \text{lbf/ft}^2\end{aligned}$$

Determine the skin-friction capacity of each layer.

For layer 1,

$$\begin{aligned}Q_{f,1} &= f_{s,1}\pi B_1 L_1 \\ &= \left(161.3\ \frac{\text{lbf}}{\text{ft}^2}\right)\pi\left(\frac{18\ \text{in}}{12\ \frac{\text{in}}{\text{ft}}}\right)(10\ \text{ft}) \\ &= 7597\ \text{lbf}\end{aligned}$$

For layer 2,

$$\begin{aligned}Q_{f,2} &= f_{s,2}\pi B_2 L_2 \\ &= \left(362.3\ \frac{\text{lbf}}{\text{ft}^2}\right)\pi\left(\frac{18\ \text{in}}{12\ \frac{\text{in}}{\text{ft}}}\right)(4\ \text{ft}) \\ &= 6811\ \text{lbf}\end{aligned}$$

For layer 3,

$$\begin{aligned}Q_{f,3} &= f_{s,3}\pi B_3 L_3 \\ &= \left(520.7\ \frac{\text{lbf}}{\text{ft}^2}\right)\pi\left(\frac{18\ \text{in}}{12\ \frac{\text{in}}{\text{ft}}}\right)(16\ \text{ft}) \\ &= 39{,}261\ \text{lbf}\end{aligned}$$

For layer 4,

$$\begin{aligned}Q_{f,4} &= f_{s,4}\pi B_4 L_4 \\ &= \left(715.6\ \frac{\text{lbf}}{\text{ft}^2}\right)\pi\left(\frac{18\ \text{in}}{12\ \frac{\text{in}}{\text{ft}}}\right)(10\ \text{ft}) \\ &= 33{,}705\ \text{lbf}\end{aligned}$$

The total skin-friction capacity is

$$\begin{aligned}Q_{f,t} &= 7597\ \text{lbf} + 6811\ \text{lbf} + 39{,}261\ \text{lbf} + 33{,}705\ \text{lbf} \\ &= 87{,}374\ \text{lbf}\end{aligned}$$

The end bearing is located in layer 4. The value for cohesion is typically half the unconfined compressive strength value. Calculate the total pile tip capacity.

$$\begin{aligned}Q_p &= 9cA_p = (9)\left(\frac{\left(14\ \frac{\text{lbf}}{\text{in}^2}\right)\left(12\ \frac{\text{in}}{\text{ft}}\right)^2}{2}\right)\left(\frac{\pi}{4}\right)\left(\frac{18\ \text{in}}{12\ \frac{\text{in}}{\text{ft}}}\right)^2 \\ &= 16{,}032\ \text{lbf}\end{aligned}$$

The total capacity of the pile is

$$Q_{\text{ult}} = Q_p + Q_f = \frac{16{,}032 \text{ lbf} + 87{,}374 \text{ lbf}}{1000 \ \frac{\text{lbf}}{\text{kip}}}$$
$$= 103.406 \text{ kips} \quad (100 \text{ kips})$$

Author Commentary

The average effective stress is commonly taken midway through a soil layer.

The answer is (D).

43. Multiply the modulus of subgrade reaction by the pier diameter to determine the maximum allowable soil reaction. The soil reaction must be per unit length of pier, so assume a value of 1 in for the deflection, Δ.

$$\begin{aligned} p_{\text{max}} &= kD\Delta \\ &= \left(140 \ \frac{\text{lbf}}{\text{in}^3}\right)(42 \text{ in})(1 \text{ in}) \\ &= 5880 \text{ lbf/in} \quad (5900 \text{ lbf/in}) \end{aligned}$$

Author Commentary

The use of a unit value, or 1, is required when determining values such as a unit length. If this value is not added, the units will not work out correctly.

The answer is (D).

44. Use the Terzaghi-Meyerhof equation for ultimate bearing capacity. Modify the unit weight for buoyancy to account for the groundwater location.

$$q_{\text{ult}} = \frac{1}{2}\gamma_b B N_\gamma + (p_q + \gamma D_f) N_q$$

From Terzaghi factor tables, for $\phi = 30°$, the bearing capacity factors N_γ and N_q are 15.7 and 18.4, respectively.

Since the overburden correction can be ignored, the net bearing capacity need not be calculated. Substitute the ultimate bearing capacity equation into the equation for allowable bearing capacity, and solve in terms of the footing width, B.

$$\begin{aligned} q_a &= \frac{q_{\text{ult}}}{F} = \frac{\frac{1}{2}\gamma_b B N_\gamma + (p_q + \gamma D_f) N_q}{F} \\ &= \frac{\left(\frac{1}{2}\right)(\gamma - \gamma_w) B N_\gamma + (p_q + \gamma D_f) N_q}{F} \\ &= \frac{\left(\frac{1}{2}\right)\left(117.7 \ \frac{\text{lbf}}{\text{ft}^3} - 62.4 \ \frac{\text{lbf}}{\text{ft}^3}\right)B(15.7) + \left(0 + \left(117.7 \ \frac{\text{lbf}}{\text{ft}^3}\right)\left(\frac{18 \text{ in}}{12 \ \frac{\text{in}}{\text{ft}}}\right)\right)(18.4)}{1.5} \\ &= \left(289.4 \ \frac{\text{lbf}}{\text{ft}^3}\right)B + 2165.7 \ \frac{\text{lbf}}{\text{ft}^2} \end{aligned}$$

The actual contact pressure, expressed in terms of B, is

$$\begin{aligned} p_{\text{actual}} &= \frac{p_c}{B} \\ &= \frac{\left(30 \ \frac{\text{kips}}{\text{ft}}\right)\left(1000 \ \frac{\text{lbf}}{\text{kip}}\right)}{B} \\ &= \frac{30{,}000 \ \frac{\text{lbf}}{\text{ft}}}{B} \end{aligned}$$

Compare the actual contact pressure due to the load conditions with the predicted contact pressure as determined by the bearing capacity equation and given factor of safety.

$$\begin{aligned} p_{\text{actual}} &= q_a \\ \frac{30{,}000 \ \frac{\text{lbf}}{\text{ft}}}{B} &= \left(289.4 \ \frac{\text{lbf}}{\text{ft}^3}\right)B + 2165.7 \ \frac{\text{lbf}}{\text{ft}^2} \\ 0 &= \left(289.4 \ \frac{\text{lbf}}{\text{ft}^3}\right)B^2 + \left(2165.7 \ \frac{\text{lbf}}{\text{ft}^2}\right)B \\ &\quad -30{,}000 \ \frac{\text{lbf}}{\text{ft}} \end{aligned}$$

The equation is a second-degree polynomial, so the quadratic formula may be used to solve for the required width.

$$x_1, x_2 = \frac{-b \pm \sqrt{b^2 - 4ac}}{2a}$$

Using substitution, $a = 289.4$ lbf/ft^3, $b = 2165.7$ lbf/ft^2, and $c = -30{,}000$ lbf/ft. Calculate the two roots, x_1 and x_2.

$$
\begin{aligned}
x_1 &= \frac{-b + \sqrt{b^2 - 4ac}}{2a} \\
&= \frac{-2165.7\ \frac{\text{lbf}}{\text{ft}^2} + \sqrt{\begin{array}{l}\left(2165.7\ \frac{\text{lbf}}{\text{ft}^2}\right)^2 - (4) \\ \times\left(289.4\ \frac{\text{lbf}}{\text{ft}^3}\right) \\ \times\left(-30{,}000\ \frac{\text{lbf}}{\text{ft}}\right)\end{array}}}{(2)\left(289.4\ \frac{\text{lbf}}{\text{ft}^3}\right)} \\
&= 7.1\ \text{ft}
\end{aligned}
$$

$$
\begin{aligned}
x_2 &= \frac{-b - \sqrt{b^2 - 4ac}}{2a} \\
&= \frac{-2165.7\ \frac{\text{lbf}}{\text{ft}^2} - \sqrt{\begin{array}{l}\left(2165.7\ \frac{\text{lbf}}{\text{ft}^2}\right)^2 - (4) \\ \times\left(289.4\ \frac{\text{lbf}}{\text{ft}^3}\right) \\ \times\left(-30{,}000\ \frac{\text{lbf}}{\text{ft}}\right)\end{array}}}{(2)\left(289.4\ \frac{\text{lbf}}{\text{ft}^3}\right)} \\
&= -14.6\ \text{ft} \quad \text{[disregard this root]}
\end{aligned}
$$

The required width of the footing is 7.1 ft.

Author Commentary

Using the quadratic formula instead of using trial and error can save time on the exam. Since the constant term of the quadratic solution is negative and the first root, x_1, is positive, x_2 does not need to be calculated, as it will be negative.

The answer is (D).

45. To use the SPT result to determine the allowable soil pressure, the SPT result must be corrected for the presence of groundwater within the zone of influence.

$$q_a = 0.22 C_n N C_w$$

Since the unit weight of the soil is close to 100 lbf/ft^3, the SPT correction factor, C_n, can be taken as 1. The correction for groundwater is calculated with consideration for the width of the proposed raft foundation.

$$
\begin{aligned}
C_w &= 0.5 + 0.5\left(\frac{D_w}{D_f + B}\right) \\
&= 0.5 + (0.5)\left(\frac{10\ \text{ft}}{5\ \text{ft} + 35\ \text{ft}}\right) \\
&= 0.625
\end{aligned}
$$

The allowable soil pressure is

$$
\begin{aligned}
q_a &= 0.22 C_n N C_w = (0.22)(1)(8)(0.625) \\
&= 1.1\ \text{tons/ft}^2
\end{aligned}
$$

Author Commentary

Tab your reference material to quickly find the correction factors for overburden and the presence of a water table.

The answer is (C).

46. A fully compensated raft foundation can support a structural load equivalent to the weight of the excavated soil. Using this relationship, determine the required bearing depth.

$$\frac{P_{\text{load}}}{A_{\text{raft}}} = \gamma D_f = \gamma_{\text{dry}}(1 + w) D_f$$

$$
\begin{aligned}
D_f &= \frac{\frac{P_{\text{load}}}{A_{\text{raft}}}}{\gamma_{\text{dry}}(1 + w)} = \frac{\frac{(2250\ \text{tons})\left(2000\ \frac{\text{lbf}}{\text{ton}}\right)}{(50\ \text{ft})(75\ \text{ft})}}{\left(107.9\ \frac{\text{lbf}}{\text{ft}^3}\right)(1 + 0.12)} \\
&= 9.9\ \text{ft} \quad (10\ \text{ft})
\end{aligned}
$$

Author Commentary

The term "fully compensated" means that the weight of the excavated soil is equivalent to the structural load.

The answer is (B).

47. The total pile capacity, or ultimate static capacity, is the sum of the point-bearing capacity, or tip capacity, and the skin-friction capacity, or side capacity.

$$Q_{\text{ult}} = Q_p + Q_f$$

Use the α-method, and calculate the skin-friction capacity, Q_f, and skin-friction factor, f_s.

$$Q_f = f_s 4BL$$
$$f_s = \alpha c$$

Determine the skin-friction factor for each soil layer. The adhesion factor is the ratio of the skin-friction factor and the cohesion. The value for cohesion is typically half the unconfined compressive strength.

layer	unconfined compressive strength, q_u (lbf/in^2)	cohesion, c (lbf/in^2)	cohesion, c (lbf/ft^2)	adhesion factor, α	skin friction factor, f_s (lbf/ft^2)
1	–	0.7	101	1	101
2	–	0.7	101	1	101
3	9	4.5	648	1	648
4	28	14	2016	0.56	1129

To calculate the skin-friction capacity, first determine the surface area of the pile passing through each soil layer.

For layer 1,

$$\begin{aligned} A_{s,1} &= 4B_1L_1 = (4)(1\ \text{ft})(6\ \text{ft}) \\ &= 24\ \text{ft}^2 \end{aligned}$$

For layer 2,

$$\begin{aligned} A_{s,2} &= 4B_2L_2 = (4)(1\ \text{ft})(8\ \text{ft}) \\ &= 32\ \text{ft}^2 \end{aligned}$$

For layer 3,

$$\begin{aligned} A_{s,3} &= 4B_3L_3 = (4)(1\ \text{ft})(11\ \text{ft}) \\ &= 44\ \text{ft}^2 \end{aligned}$$

For layer 4,

$$\begin{aligned} A_{s,4} &= 4B_4L_4 = (4)(1\ \text{ft})(10\ \text{ft}) \\ &= 40\ \text{ft}^2 \end{aligned}$$

The total skin-friction capacity of each layer is the product of the skin friction factor and the surface area of the pile passing through that layer.

For layer 1,

$$\begin{aligned} Q_{f,1} &= f_{s,1}A_{s,1} = \left(101\ \frac{\text{lbf}}{\text{ft}^2}\right)(24\ \text{ft}^2) \\ &= 2424\ \text{lbf} \end{aligned}$$

For layer 2,

$$\begin{aligned} Q_{f,2} &= f_{s,2}A_{s,2} = \left(101\ \frac{\text{lbf}}{\text{ft}^2}\right)(32\ \text{ft}^2) \\ &= 3232\ \text{lbf} \end{aligned}$$

For layer 3,

$$\begin{aligned} Q_{f,3} &= f_{s,3}A_{s,3} = \left(648\ \frac{\text{lbf}}{\text{ft}^2}\right)(44\ \text{ft}^2) \\ &= 28{,}512\ \text{lbf} \end{aligned}$$

For layer 4,

$$\begin{aligned} Q_{f,4} &= f_{s,4}A_{s,4} = \left(1129\ \frac{\text{lbf}}{\text{ft}^2}\right)(40\ \text{ft}^2) \\ &= 45{,}160\ \text{lbf} \end{aligned}$$

The total skin-friction capacity is

$$\begin{aligned} Q_{f,t} &= 2424\ \text{lbf} + 3232\ \text{lbf} + 28{,}512\ \text{lbf} + 45{,}160\ \text{lbf} \\ &= 79{,}328\ \text{lbf} \end{aligned}$$

The end bearing is located in layer 4. Calculate the point-bearing capacity.

$$\begin{aligned} Q_p &= 9cA_p \\ &= \frac{(9)\left(2016\ \frac{\text{lbf}}{\text{ft}^2}\right)(12\ \text{in})(12\ \text{in})}{\left(12\ \frac{\text{in}}{\text{ft}}\right)^2} \\ &= 18{,}144\ \text{lbf} \end{aligned}$$

The total pile capacity is

$$\begin{aligned} Q_{\text{ult}} &= Q_f + Q_p \\ &= \frac{79{,}328\ \text{lbf} + 18{,}144\ \text{lbf}}{1000\ \frac{\text{lbf}}{\text{kip}}} \\ &= 97.472\ \text{kips} \quad (97\ \text{kips}) \end{aligned}$$

Author Commentary

Solutions to deep foundation problems might involve repetitive calculations, such as those for surface areas and friction factors. These can be saved in your calculator's memory and easily tabulated.

The answer is (B).

48. Since a unit weight for the clay is not given, the ultimate bearing capacity and the net bearing capacity are assumed to be equivalent.

$$q_{\text{ult}} = q_{\text{net}} = cN_c$$

The clay is saturated and may be assumed to remain undrained due to low permeability. Use the angle of internal friction, ϕ, corresponding to undrained conditions to find the Terzaghi bearing capacity factor, N_c. For undrained conditions, the angle of internal friction is typically taken as zero. For $\phi = 0$, N_c is 5.7. Calculate the net bearing capacity.

$$\begin{aligned} q_{\text{net}} &= cN_c = \left(150\ \frac{\text{lbf}}{\text{ft}^2}\right)(5.7) \\ &= 855\ \text{lbf/ft}^2 \end{aligned}$$

Adjusting for a factor of safety of 3, the allowable bearing pressure is

$$\begin{aligned} q_a &= \frac{q_{\text{net}}}{F} = \frac{855\ \frac{\text{lbf}}{\text{ft}^2}}{3} \\ &= 285\ \text{lbf/ft}^2 \quad (280\ \text{lbf/ft}^2) \end{aligned}$$

Author Commentary

Look for ways to simplify equations. In this problem, since the unit weight of the clay is not given, two terms may be omitted from the equation.

The answer is (B).

49. The failure plane can be considered to be the surface area of the pile. Using the Mohr-Coulomb equation, determine the shear strength of the soil in response to normal stress along the failure plane. This can be expressed along with the coefficient of lateral earth pressure to calculate the skin friction.

$$f_s = c_A + k_s(\sigma_v \tan\delta)$$

Therefore, the skin-friction capacity for an individual layer is

$$Q_f = A_s f_s = \pi DL\big(c_A + k(\sigma_v \tan\delta)\big)$$

Using effective stress analysis where necessary, determine the overburden (confining) stress at the midpoint of each soil layer.

For layer 1,

$$\begin{aligned} \sigma'_{v1,\text{ave}} &= \gamma_{\text{dry},1}\left(\frac{L_1}{2}\right) \\ &= \left(108.5\ \frac{\text{lbf}}{\text{ft}^3}\right)\left(\frac{6\ \text{ft}}{2}\right) \\ &= 326\ \text{lbf/ft}^2 \end{aligned}$$

For layer 2,

$$\begin{aligned} \sigma'_{v2,\text{ave}} &= \gamma_{\text{dry},2}L_1 + (\gamma_{\text{dry},2}(1+w_2) - \gamma_w)\left(\frac{L_2}{2}\right) \\ &= \left(108.5\ \frac{\text{lbf}}{\text{ft}^3}\right)(6\ \text{ft}) \\ &\quad + \left(\begin{array}{c}\left(108.5\ \frac{\text{lbf}}{\text{ft}^3}\right)(1+0.143) \\ -62.4\ \frac{\text{lbf}}{\text{ft}^3}\end{array}\right)\left(\frac{7\ \text{ft}}{2}\right) \\ &= 867\ \text{lbf/ft}^2 \end{aligned}$$

For layer 3,

$$\begin{aligned} \sigma'_{v3,\text{ave}} &= \gamma_{\text{dry},3}L_1 + (\gamma_{\text{dry},3}(1+w_2) - \gamma_w)L_2 \\ &\quad + (\gamma_{\text{sat},3} - \gamma_w)\left(\frac{L_3}{2}\right) \\ &= \left(108.5\ \frac{\text{lbf}}{\text{ft}^3}\right)(6\ \text{ft}) + \left(\begin{array}{c}\left(108.5\ \frac{\text{lbf}}{\text{ft}^3}\right)(1+0.143) \\ -62.4\ \frac{\text{lbf}}{\text{ft}^3}\end{array}\right) \\ &\quad \times(7\ \text{ft}) + \left(116.4\ \frac{\text{lbf}}{\text{ft}^3} - 62.4\ \frac{\text{lbf}}{\text{ft}^3}\right)\left(\frac{10\ \text{ft}}{2}\right) \\ &= 1352\ \text{lbf/ft}^2 \end{aligned}$$

The pile penetrates into layer 4 by

$$L_4 = 28\ \text{ft} - (6\ \text{ft} + 7\ \text{ft} + 10\ \text{ft}) = 5\ \text{ft}$$

For layer 4,

$$\begin{aligned}\sigma'_{v4,\text{ave}} &= \gamma_{\text{dry},1}L_1 + \big(\gamma_{\text{dry},1}(1+w_2)-\gamma_w\big)L_2 \\ &\quad + (\gamma_{\text{sat},3}-\gamma_w)L_3 + (\gamma_{\text{sat},4}-\gamma_w)\left(\frac{L_4}{2}\right) \\ &= \left(108.5\ \frac{\text{lbf}}{\text{ft}^3}\right)(6\ \text{ft}) \\ &\quad + \left(\left(108.5\ \frac{\text{lbf}}{\text{ft}^3}\right)(1+0.143) - 62.4\ \frac{\text{lbf}}{\text{ft}^3}\right)(7\ \text{ft}) \\ &\quad + \left(116.4\ \frac{\text{lbf}}{\text{ft}^3} - 62.4\ \frac{\text{lbf}}{\text{ft}^3}\right)(10\ \text{ft}) \\ &\quad + \left(124.3\ \frac{\text{lbf}}{\text{ft}^3} - 62.4\ \frac{\text{lbf}}{\text{ft}^3}\right)\left(\frac{5\ \text{ft}}{2}\right) \\ &= 1777\ \text{lbf/ft}^2\end{aligned}$$

Since the pile is driven into clay, assume the coefficient of lateral earth pressure is 1.0. Determine the skin-friction factor for each soil layer. For smooth concrete, $0.8c \le c_A \le c$. The cohesion values for layers 1 and 2 are equal.

Determine the friction capacity for each layer. For layer 1,

$$\begin{aligned}Q_{f,1} &= \pi DL\big(c_A + k(\sigma_v \tan\delta)\big) \\ &= \pi DL_1\big(0.8c_1 + k(\sigma'_{v1,\text{ave}}\tan\delta)\big) \\ &= \pi\left(\frac{14\ \text{in}}{12\ \frac{\text{in}}{\text{ft}}}\right)(6\ \text{ft}) \\ &\quad\times\left((0.8)\left(48\ \frac{\text{lbf}}{\text{ft}^2}\right) + (1.0)\left(\left(326\ \frac{\text{lbf}}{\text{ft}^2}\right)\tan 15^\circ\right)\right) \\ &= 2762\ \text{lbf}\end{aligned}$$

For layer 2,

$$\begin{aligned}Q_{f,2} &= \pi DL_2\big(0.8c_2 + k(\sigma'_{v2,\text{ave}}\tan\delta)\big) \\ &= \pi\left(\frac{14\ \text{in}}{12\ \frac{\text{in}}{\text{ft}}}\right)(7\ \text{ft}) \\ &\quad\times\left((0.8)\left(48\ \frac{\text{lbf}}{\text{ft}^2}\right) + (1.0)\left(\left(867\ \frac{\text{lbf}}{\text{ft}^2}\right)\tan 15^\circ\right)\right) \\ &= 6943\ \text{lbf}\end{aligned}$$

For layer 3,

$$\begin{aligned}Q_{f,3} &= \pi DL_3\big(0.8c_3 + k(\sigma'_{v3,\text{ave}}\tan\delta)\big) \\ &= \pi\left(\frac{14\ \text{in}}{12\ \frac{\text{in}}{\text{ft}}}\right)(10\ \text{ft}) \\ &\quad\times\left((0.8)\left(400\ \frac{\text{lbf}}{\text{ft}^2}\right) + (1.0)\left(\left(1352\ \frac{\text{lbf}}{\text{ft}^2}\right)\tan 8^\circ\right)\right) \\ &= 18{,}694\ \text{lbf}\end{aligned}$$

The pile does not drive completely through layer 4, so the cohesion is taken as half the value of q_u, or $(1000\ \text{lbf/ft}^2)/2 = 500\ \text{lbf/ft}^2$. For layer 4,

$$\begin{aligned}Q_{f,4} &= \pi DL_4\big(0.8c_4 + k(\sigma'_{v4,\text{ave}}\tan\delta)\big) \\ &= \pi\left(\frac{14\ \text{in}}{12\ \frac{\text{in}}{\text{ft}}}\right)(5\ \text{ft}) \\ &\quad\times\left((0.8)\left(500\ \frac{\text{lbf}}{\text{ft}^2}\right) + (1.0)\left(\left(1777\frac{\text{lbf}}{\text{ft}^2}\right)\tan 18^\circ\right)\right) \\ &= 17{,}912\ \text{lbf}\end{aligned}$$

Determine the skin-friction capacity of the entire pile length by summing the friction capacity for each layer.

$$\begin{aligned} Q_f &= \sum Q_{f,i} = Q_{f,1} + Q_{f,2} + Q_{f,3} + Q_{f,4} \\ &= \frac{2762 \text{ lbf} + 6943 \text{ lbf} + 18{,}694 \text{ lbf} + 17{,}912 \text{ lbf}}{1000 \ \frac{\text{lbf}}{\text{kip}}} \\ &= 46.3 \text{ kips} \quad (46 \text{ kips}) \end{aligned}$$

Author Commentary

Remember to account for the fact that the pile does not drive completely through layer 4.

The answer is (B).

50. Use the Engineer News Record (ENR) equation for a drop hammer to determine the allowable pile capacity. The distance S remains in units of inches.

$$\begin{aligned} Q_{a,\text{lbf}} &= \frac{Q_{\text{ult}}}{FS} \\ &= \frac{2W_{\text{hammer,lbf}} H_{\text{fall,ft}}}{S_{\text{in}} + 1} \\ &= \frac{(2)(10{,}000 \text{ lbf})\left(\frac{36 \text{ in}}{12 \ \frac{\text{in}}{\text{ft}}}\right)}{(0.75 \text{ in} + 1)\left(1000 \ \frac{\text{lbf}}{\text{kip}}\right)} \\ &= 34.3 \text{ kips} \quad (34 \text{ kips}) \end{aligned}$$

Author Commentary

Be sure to understand that certain terms in an equation may require specific units. In this problem, the penetration per blow is in inches, while the drop distance is in feet.

The answer is (A).

51. Using the Terzaghi-Meyerhof equation for ultimate bearing capacity, solve for the diameter using the appropriate shape factors, S_γ and S_c.

$$q_{\text{ult}} = \frac{1}{2}\gamma B N_\gamma S_\gamma + c N_c S_c + (p_q + \gamma D_f) N_q$$

Use Terzaghi factor and shape factor tables to find the Terzaghi factors, N_c, N_γ, and N_q, for $\phi = 25°$, and the shape factors, S_γ and S_c, for a circular footing.

$$\begin{aligned} N_\gamma &= 9.7 \\ N_c &= 25.1 \\ N_q &= 12.7 \\ S_\gamma &= 0.70 \\ S_c &= 1.20 \end{aligned}$$

Since overburden correction can be ignored, the net bearing capacity need not be calculated. Substituting the ultimate bearing capacity equation into the equation for allowable bearing capacity, the allowable bearing capacity in terms of the footing width, B, is

$$\begin{aligned} q_a &= \frac{q_{\text{ult}}}{F} = \frac{\frac{1}{2}\gamma B N_\gamma S_\gamma + c N_c S_c + (p_q + \gamma D_f) N_q}{F} \\ &= \frac{\left(\frac{1}{2}\right)\left(112.5 \ \frac{\text{lbf}}{\text{ft}^3}\right) B (9.7)(0.70) + \left(300 \ \frac{\text{lbf}}{\text{ft}^2}\right)(25.1)(1.20) + \left(\left(112.5 \ \frac{\text{lbf}}{\text{ft}^3}\right)\left(\frac{36 \text{ in}}{12 \ \frac{\text{in}}{\text{ft}}}\right)\right)(12.7)}{2.5} \\ &= \left(152.8 \ \frac{\text{lbf}}{\text{ft}^3}\right) B + 5328.9 \ \frac{\text{lbf}}{\text{ft}^2} \end{aligned}$$

Using the diameter and area of the circular footing, the actual contact pressure is

$$\begin{aligned} p_{\text{actual}} &= \frac{P_c}{A_f} = \frac{P}{\frac{1}{4}\pi B^2} \\ &= \frac{(28 \text{ kips})\left(1000 \ \frac{\text{lbf}}{\text{kip}}\right)}{\frac{1}{4}\pi B^2} \\ &= \frac{28{,}000 \text{ lbf}}{0.785 B^2} \end{aligned}$$

Compare the actual contact pressure due to the load conditions with the predicted contact pressure as determined by the bearing capacity equation and given factor of safety.

$$p_{\text{actual}} = q_a$$

$$\frac{28{,}000\ \text{lbf}}{0.785B^2} = \left(152.8\ \frac{\text{lbf}}{\text{ft}^3}\right)B + 5328.9\ \frac{\text{lbf}}{\text{ft}^2}$$

$$28{,}000\ \text{lbf} = \left(120\ \frac{\text{lbf}}{\text{ft}^3}\right)B^3 + \left(4185.3\ \frac{\text{lbf}}{\text{ft}^2}\right)B^2$$

Solve for the diameter by trial and error. Starting with the common footing diameter of 2 ft, calculate a preliminary answer for comparison with a larger value, such as 3 ft.

For $B = 2$ ft,

$$\left(120\ \frac{\text{lbf}}{\text{ft}^3}\right)(2\ \text{ft})^3 + \left(4185.3\ \frac{\text{lbf}}{\text{ft}^2}\right)(2\ \text{ft})^2 = 17{,}701\ \text{lbf} \quad [<28{,}000\ \text{lbf}]$$

For $B = 3$ ft,

$$\left(120\ \frac{\text{lbf}}{\text{ft}^3}\right)(3\ \text{ft})^3 + \left(4185.3\ \frac{\text{lbf}}{\text{ft}^2}\right)(3\ \text{ft})^2 = 40{,}908\ \text{lbf} \quad [>28{,}000\ \text{lbf}]$$

Since a diameter of 2 ft results in a value less than the required capacity and a diameter of 3 ft results in a value much greater than the required capacity, choose a value between 2 ft and 3 ft for the next calculation.

For $B = 2.5$ ft,

$$\left(120\ \frac{\text{lbf}}{\text{ft}^3}\right)(2.5\ \text{ft})^3 + \left(4185.3\ \frac{\text{lbf}}{\text{ft}^2}\right)(2.5\ \text{ft})^2 = 28{,}033\ \text{lbf} \quad [>28{,}000\ \text{lbf}]$$

A diameter of 2.5 ft results in a capacity that exceeds the required capacity and is therefore suitable. For most footing design applications, dimensions within 6 in are typically suitable.

Author Commentary

The best approach to solving this problem is through trial and error. If the equation were in the form of a second-degree polynomial, using the quadratic formula would be a quicker approach to solving the problem.

The answer is (C).

52. Since Young's modulus varies linearly with depth, the modulus of subgrade reaction is similar throughout the length of the pile and is independent of depth. Determine the modulus of subgrade reaction by finding the slope of the given data.

$$k = \frac{\Delta E}{\Delta D} = \frac{1800\ \frac{\text{kips}}{\text{ft}^2} - 600\ \frac{\text{kips}}{\text{ft}^2}}{15\ \text{ft} - 5\ \text{ft}} = 120\ \text{kips/ft}^3$$

Multiply the modulus of subgrade reaction by the pile diameter and the given lateral deflection of 1 in to find the soil reaction.

$$p = kd\Delta = \left(\frac{\left(120\ \frac{\text{kips}}{\text{ft}^3}\right)(18\ \text{in})(1\ \text{in})}{\left(12\ \frac{\text{in}}{\text{ft}}\right)^3}\right)\left(1000\ \frac{\text{lbf}}{\text{kip}}\right) = 1250\ \text{lbf/in} \quad (1300\ \text{lbf/in})$$

Author Commentary

Young's modulus and the modulus of elasticity are essentially equal, and can be treated as such.

The answer is (C).

53. The maximum soil reaction is

$$\frac{p}{p_{\max}} \times 100\% = \frac{300\ \frac{\text{lbf}}{\text{in}}}{650\ \frac{\text{lbf}}{\text{in}}} \times 100\% = 46\%$$

Author Commentary

A p-y curve is a graph of soil reaction and vertical depth. p-y curves are common for determining lateral pile capacity in deep foundation evaluations.

The answer is (D).

54. Compare the skin-friction capacity calculated from the combined surface area of all piles with that based on the perimeter of the pile group as a block. The lower value controls.

Use the α-method to calculate the skin-friction capacity of a single pile. The cohesion is 500 lbf/ft^2, so assume a standard adhesion factor value of 1.0.

$$Q_f = A_s f_s = \pi BL\alpha c = \pi\left(\frac{30\ \text{in}}{12\ \frac{\text{in}}{\text{ft}}}\right)(20\ \text{ft})(1.0)\left(500\ \frac{\text{lbf}}{\text{ft}^2}\right) = 78{,}540\ \text{lbf}$$

Find the skin-friction capacity of all piles combined.

$$\begin{aligned}\sum Q_f &= N_{\text{piles}} Q_f \\ &= \frac{(12)(78{,}450 \text{ lbf})}{1000 \ \frac{\text{lbf}}{\text{kip}}} \\ &= 942 \text{ kips}\end{aligned}$$

Calculate the total skin-friction (side) capacity of the pile group as a block.

$$Q_{s,\text{block}} = 2(b + w)L_e c$$

The base length of the pile group is

$$\begin{aligned}b &= \frac{(3)(78 \text{ in}) + 30 \text{ in}}{12 \ \frac{\text{in}}{\text{ft}}} \\ &= 22 \text{ ft}\end{aligned}$$

The base width of the pile group is

$$\begin{aligned}w &= \frac{(2)(78 \text{ in}) + 30 \text{ in}}{12 \ \frac{\text{in}}{\text{ft}}} \\ &= 15.5 \text{ ft}\end{aligned}$$

The total skin-friction capacity of the pile group as a block is

$$\begin{aligned}Q_{s,\text{block}} &= 2(b + w)L_e c \\ &= \frac{(2)(22 \text{ ft} + 15.5 \text{ ft})(20 \text{ ft})\left(500 \ \frac{\text{lbf}}{\text{ft}^2}\right)}{1000 \ \frac{\text{lbf}}{\text{kip}}} \\ &= 750 \text{ kips}\end{aligned}$$

The skin-friction capacity of the pile group as a block is lower and should be chosen instead of the individual capacities combined.

Author Commentary

A licensed and practicing engineer must be able to recognize which capacity value is safest to use. The lower capacity will fail first. Therefore, the lower capacity value should be chosen for design.

The answer is (B).

55. If the combined circumference of the individual piles is less than the circumference of the two piles as a group, they can be considered to act individually under loading.

Determine the circumference of the pile group as a block.

$$\begin{aligned}C_{\text{block}} &= \pi d + 2(\text{spacing}) \\ &= \pi(30 \text{ in}) + (2)(42 \text{ in}) \\ &= 178.25 \text{ in}\end{aligned}$$

Determine the circumference of an individual pile.

$$\begin{aligned}C_{\text{indiv}} &= \pi d = \pi(30 \text{ in}) \\ &= 94.25 \text{ in}\end{aligned}$$

Combined for the two piles,

$$94.24 \text{ in} + 94.24 \text{ in} = 188.45 \text{ in}$$

The circumference of the piles as a group is less than the combined circumference of the individual piles. Therefore, the two piles will act as a group.

Author Commentary

If the structural loading and dimensions for both piles are similar, and both are installed in the same soil profile, the pile action is determined based on a simple comparison of the circumference of the individual piles added together or taken as a block. The larger of the two circumferences requires more soil mobilization. Therefore, the smaller circumference will govern.

The answer is (B).

56. Calculate the resistance to sliding. Because the backfill is level and the wall has a straight face, there is no vertical resultant force. The resistance to sliding is

$$\begin{aligned}R_{\text{SL}} &= \left(\sum W_i + R_{a,v}\right)\tan\phi + c_A B \\ &= \left(\sum \gamma_i A_i + R_{a,v}\right)\tan\phi + c_A B \\ &= \left(\left(145 \ \frac{\text{lbf}}{\text{ft}^3}\right)(12 \text{ ft})(6 \text{ ft}) + 0 \ \frac{\text{lbf}}{\text{ft}}\right)\tan 30^\circ \\ &\quad + \left(300 \ \frac{\text{lbf}}{\text{ft}^2}\right)(6 \text{ ft}) \\ &= 7828 \text{ lbf/ft} \quad (7800 \text{ lbf/ft})\end{aligned}$$

Author Commentary

Be sure to use the correct equation for resistance to sliding. The applicable equation depends on the presence or absence of a key.

The answer is (D).

57. The equation for consolidation settlement is

$$S = C_c\left(\frac{H_o}{1+e_o}\right)\log_{10}\frac{p_o' - \Delta p_{f,\mathrm{wt}}' + \Delta p_v'}{p_o'}$$

Use the 60° method to calculate the consolidation in the clay layer. Evaluate the pressure distribution from a point ⅔ the length of the pile group. Settlement above ⅔ L is assumed to be negligible, and below ⅔L, the pressure distribution spreads out at an angle of approximately 60° from the horizontal.

Calculate the width of the pile group.

$$\begin{aligned} B &= (N_{\mathrm{piles,base}} - 1)d + D \\ &= \frac{(4-1)(90 \text{ in}) + 30 \text{ in}}{12 \, \frac{\text{in}}{\text{ft}}} \\ &= 25 \text{ ft} \end{aligned}$$

Calculate the length of the pile group.

$$\begin{aligned} L &= (N_{\mathrm{piles,length}} - 1)d + D \\ &= \frac{(4-1)(90 \text{ in}) + 30 \text{ in}}{12 \, \frac{\text{in}}{\text{ft}}} \\ &= 25 \text{ ft} \end{aligned}$$

Determine the area of influence at the midpoint of the soft clay layer, 16 ft below ⅔L.

$$\begin{aligned} A &= (B + 2(h \cot 60°))(L + 2(h \cot 60°)) \\ &= (25 \text{ ft} + (2)((16 \text{ ft})\cot 60°)) \\ &\quad \times (25 \text{ ft} + (2)((16 \text{ ft})\cot 60°)) \\ &= 1891 \text{ ft}^2 \end{aligned}$$

The ultimate pile capacity is the sum of the point-bearing (tip) capacity and the skin-friction (side) capacity.

$$Q_{\mathrm{ult}} = Q_p + Q_f$$

To find the side-friction capacity, first use the α-method to determine the skin-friction factor for an individual pile in the clayey silt. Typically, the adhesion value for a cohesion of approximately 1000 lbf/ft^2 is from 0.56 to 0.96, with the average value being 0.83. Use the average value of 0.83.

$$\begin{aligned} f_s &= \alpha c \\ &= (0.83)\left(7.0 \, \frac{\text{lbf}}{\text{in}^2}\right)\left(12 \, \frac{\text{in}}{\text{ft}}\right)^2 \\ &= 837 \text{ lbf/ft}^2 \end{aligned}$$

Calculate the surface area of the pile group passing through the clayey silt layer.

$$\begin{aligned} A_s &= \pi D L \\ &= \pi\left(\frac{30 \text{ in}}{12 \, \frac{\text{in}}{\text{ft}}}\right)(27 \text{ ft}) \\ &= 212 \text{ ft}^2 \end{aligned}$$

Calculate the total skin-friction capacity for 16 piles.

$$\begin{aligned} Q_f &= A_s f_s N_{\mathrm{piles}} \\ &= (212 \text{ ft}^2)\left(837 \, \frac{\text{lbf}}{\text{ft}^2}\right)(16) \\ &= 2{,}839{,}104 \text{ lbf} \end{aligned}$$

Typically, the portion of the pile group extending below the ⅔L point (i.e., the end bearing) is disregarded. However, in this case, the load is required to be based on 30% of the ultimate capacity, which is based on the sum of the individual pile capacities, and not the pile group capacity. Therefore, the point-bearing must be calculated as well. The point-bearing location is in the clayey silt. Calculate the total pile tip capacity for 16 piles.

$$\begin{aligned} Q_p &= 9cA_p = 9c\frac{\pi}{4}D^2 N_{\mathrm{piles}} \\ &= (9)\left(7.0 \, \frac{\text{lbf}}{\text{in}^2}\right)\left(12 \, \frac{\text{in}}{\text{ft}}\right)^2\left(\frac{\pi}{4}\right)\left(\frac{30 \text{ in}}{12 \, \frac{\text{in}}{\text{ft}}}\right)^2(16) \\ &= 712{,}513 \text{ lbf} \end{aligned}$$

The ultimate pile capacity is

$$\begin{aligned} Q_{\mathrm{ult}} &= Q_p + Q_f \\ &= 712{,}513 \text{ lbf} + 2{,}839{,}104 \text{ lbf} \\ &= 3{,}551{,}617 \text{ lbf} \end{aligned}$$

The fraction of the ultimate pile capacity to be loaded is

$$\begin{aligned} P_{\mathrm{pile\,group}} &= 0.30 Q_{\mathrm{ult}} \\ &= (0.30)(3{,}551{,}617 \text{ lbf}) \\ &= 1{,}065{,}485 \text{ lbf} \end{aligned}$$

The change in effective pressure due to the pile group load is

$$\begin{aligned}\Delta p_v' &= \frac{P_{\text{pile group}}}{A} \\ &= \frac{1{,}065{,}485 \text{ lbf}}{1891 \text{ ft}^2} \\ &= 563 \text{ lbf/ft}^2\end{aligned}$$

Calculate the initial effective pressure at the midpoint of the soft clay layer (layer 3).

$$\begin{aligned}p_o' &= p_{\text{clay,mid}} - u_o \\ &= \gamma_{\text{dry,silt}} H_{\text{dry silt}} + \gamma_{\text{dry,silt}}(1+w) H_{\text{wet silt}} \\ &\quad + \gamma_{\text{sat,clay}} H_{\text{clay,mid}} - \gamma_w H_w \\ &= \left(112.3 \ \frac{\text{lbf}}{\text{ft}^3}\right)(16 \text{ ft}) + \left(112.3 \ \frac{\text{lbf}}{\text{ft}^3}\right) \\ &\quad \times (1+0.18)(14 \text{ ft}) + \left(104.8 \ \frac{\text{lbf}}{\text{ft}^3}\right)\left(\frac{8 \text{ ft}}{2}\right) \\ &\quad - \left(62.4 \ \frac{\text{lbf}}{\text{ft}^3}\right)\left(14 \text{ ft} + \frac{8 \text{ ft}}{2}\right) \\ &= 2948 \text{ lbf/ft}^2\end{aligned}$$

Determine the final effective pressure that occurs at the midpoint of the clay layer due to the increase in groundwater table elevation. Due to the rise in water table elevation, the new values for the height of the dry silt and wet silt, respectively, are 6 ft and 24 ft.

$$\begin{aligned}p_{f,\text{wt}}' &= p_{\text{clay,mid}} - u_f \\ &= \gamma_{\text{dry,silt}} H_{\text{dry silt}} + \gamma_{\text{dry,silt}}(1+w) H_{\text{wet silt}} \\ &\quad + \gamma_{\text{sat,clay}} H_{\text{clay,mid}} - \gamma_w H_w \\ &= \left(112.3 \ \frac{\text{lbf}}{\text{ft}^3}\right)(6 \text{ ft}) + \left(112.3 \ \frac{\text{lbf}}{\text{ft}^3}\right) \\ &\quad \times (1+0.18)(24 \text{ ft}) + \left(104.8 \ \frac{\text{lbf}}{\text{ft}^3}\right)\left(\frac{8 \text{ ft}}{2}\right) \\ &\quad - \left(62.4 \ \frac{\text{lbf}}{\text{ft}^3}\right)\left(14 \text{ ft} + 10 \text{ ft} + \frac{8 \text{ ft}}{2}\right) \\ &= 2526 \text{ lbf/ft}^2\end{aligned}$$

The change in effective pressure due to the increase in groundwater table elevation is

$$\begin{aligned}\Delta p_{f,\text{wt}}' &= p_o' - p_{f,\text{wt}}' \\ &= 2948 \ \frac{\text{lbf}}{\text{ft}^2} - 2526 \ \frac{\text{lbf}}{\text{ft}^2} \\ &= 422 \text{ lbf/ft}^2\end{aligned}$$

Determine the total expected settlement.

$$\begin{aligned}S &= C_c\left(\frac{H_o}{1+e_o}\right)\log_{10}\frac{p_o' - \Delta p_{f,\text{wt}}' + \Delta p_v'}{p_o'} \\ &= (0.72)\left(\frac{8 \text{ ft}}{1+0.55}\right) \\ &\quad \times \left(\log_{10}\frac{2948 \ \frac{\text{lbf}}{\text{ft}^2} - 422 \ \frac{\text{lbf}}{\text{ft}^2} + 563 \ \frac{\text{lbf}}{\text{ft}^2}}{2948 \ \frac{\text{lbf}}{\text{ft}^2}}\right)\left(12 \ \frac{\text{in}}{\text{ft}}\right) \\ &= 0.91 \text{ in} \quad (1 \text{ in})\end{aligned}$$

Author Commentary

The increase in water table elevation causes a reduction in the effective pressure, while the influence of the vertical pressure from the pile group would result in an increase in the effective pressure.

The answer is (B).

58. The limiting overall settlement is 0.003 in/kip. Use the largest test load given in the table, 500 kips, to find the maximum allowable settlement.

$$\begin{aligned}S_{\max} &= P_{\text{test load,max}}\left(0.003 \ \frac{\text{in}}{\text{kip}}\right) \\ &= (500 \text{ kips})\left(0.003 \ \frac{\text{in}}{\text{kip}}\right) \\ &= 1.5 \text{ in}\end{aligned}$$

Calculate the maximum allowable pile load using the test load corresponding to the limiting settlement value of 1.5 in, 360 kips.

$$\begin{aligned}P_a &= 0.75 P_{\text{test load,lim settlement}} \\ &= (0.75)(360 \text{ kips}) \\ &= 270 \text{ kips}\end{aligned}$$

The answer is (B).

59. Calculate the coefficient of passive earth pressure for the soft clay layer.

$$\begin{aligned}k_p &= \frac{1+\sin\phi}{1-\sin\phi} = \frac{1+\sin 5^\circ}{1-\sin 5^\circ} \\ &= 1.2\end{aligned}$$

Calculate the passive earth pressure acting at the base of the footing due to the triangular pressure distribution extending beneath the excavation.

$$p_p = k_p\gamma H + 2c\sqrt{k_p}$$
$$= (1.2)\left(100.5\ \frac{\text{lbf}}{\text{ft}^3}\right)(5\ \text{ft}) + (2)\left(700\ \frac{\text{lbf}}{\text{ft}^2}\right)\sqrt{1.2}$$
$$= 2136\ \text{lbf/ft}^2 \text{ per foot of footing width}$$
$$(2100\ \text{lbf/ft}^2)$$

Author Commentary

Be sure to use the soil profile data for the soils being excavated from below the mudline, not for the backfill soils.

The answer is (D).

60. The equation for consolidation settlement is

$$S = C_c\left(\frac{H_o}{1+e_o}\right)\log_{10}\frac{p_o' + \Delta p_v'}{p_o'}$$

Calculate the initial effective pressure at the midpoint of the soft clay layer (layer 3).

$$p_o' = p_{\text{clay,mid}} - u_o$$
$$= \gamma_{\text{dry,sand}}H_{\text{dry sand}} + \gamma_{\text{dry,sand}}(1+w)H_{\text{wet sand}}$$
$$+\gamma_{\text{sat,clay}}H_{\text{clay,mid}} - \gamma_w H_w$$
$$= \left(102.6\ \frac{\text{lbf}}{\text{ft}^3}\right)(6\ \text{ft}) + \left(102.6\ \frac{\text{lbf}}{\text{ft}^3}\right)$$
$$\times(1+0.14)(20\ \text{ft}) + \left(100.2\ \frac{\text{lbf}}{\text{ft}^3}\right)\left(\frac{12\ \text{ft}}{2}\right)$$
$$-\left(62.4\ \frac{\text{lbf}}{\text{ft}^3}\right)\left(20\ \text{ft} + \frac{12\ \text{ft}}{2}\right)$$
$$= 1933.7\ \text{lbf/ft}^2$$

Use the 60° method to calculate the consolidation in the clay layer. Evaluate the pressure distribution from a point $^2/_3$ the length of the pile group. Settlement above $^2/_3$ L is assumed to be negligible, and below $^2/_3L$, the pressure distribution spreads out at an angle of approximately 60° from the horizontal.

Calculate the width of the pile group.

$$B = (N_{\text{piles,base}} - 1)d + D$$
$$= \frac{(2-1)(60\ \text{in}) + 20\ \text{in}}{12\ \frac{\text{in}}{\text{ft}}}$$
$$= 6.67\ \text{ft}$$

Calculate the length of the pile group.

$$L = (N_{\text{piles,length}} - 1)d + D$$
$$= \frac{(3-1)(60\ \text{in}) + 20\ \text{in}}{12\ \frac{\text{in}}{\text{ft}}}$$
$$= 11.67\ \text{ft}$$

Determine the area of influence at the midpoint of the clay layer, 18 ft below $^2/_3L$.

$$A = (B + 2(h\cot 60°))(L + 2(h\cot 60°))$$
$$= (6.67\ \text{ft} + (2)((18\ \text{ft})\cot 60°))$$
$$\times(11.67\ \text{ft} + (2)((18\ \text{ft})\cot 60°))$$
$$= 891\ \text{ft}^2$$

Calculate the change in effective pressure at the midpoint of the clay layer due to the increase in total load from the pile group.

$$\Delta p_v' = \frac{P_{\text{pile group}}}{A} = \left(\frac{180\ \text{kips}}{891\ \text{ft}^2}\right)\left(1000\ \frac{\text{lbf}}{\text{kip}}\right)$$
$$= 202\ \text{lbf/ft}^2$$

Determine the total expected settlement.

$$S = C_c\left(\frac{H_o}{1+e_o}\right)\log_{10}\frac{p_o' + \Delta p_v'}{p_o'}$$
$$= (0.5)\left(\frac{12\ \text{ft}}{1+0.43}\right)\left(\log_{10}\frac{1933.7\ \frac{\text{lbf}}{\text{ft}^2} + 202\ \frac{\text{lbf}}{\text{ft}^2}}{1933.7\ \frac{\text{lbf}}{\text{ft}^2}}\right)$$
$$\times\left(12\ \frac{\text{in}}{\text{ft}}\right)$$
$$= 2.2\ \text{in}\quad (2\ \text{in})$$

Author Commentary

If the $^2/_3L$ point is located within the clay layer, the midpoint of that layer is not the appropriate level to evaluate the consolidation. Instead, the midpoint between $^2/_3L$ and the base of the layer would be the correct level. Essentially, the load from a pile group is considered to be applied starting from the $^2/_3L$ point. Therefore, if that location is within a layer that can consolidate, the midpoint is based on the remaining layer thickness from $^2/_3L$ to the bottom.

The answer is (B).

61. The active resultant force is located at the centroid of a triangular distribution $H/3$ above the heel, where H is the distance from the heel to the soil surface. However, the active resultant force must be broken into vertical and horizontal components. Use active force factors for a straight slope backfill to determine the vertical and horizontal active components.

$$K_v = 5 \text{ lbf/ft}^2 \quad \text{[per foot of wall length]}$$
$$K_h = 30 \text{ lbf/ft}^2 \quad \text{[per foot of wall length]}$$

The vertical component of the active resultant force is

$$\begin{aligned} R_{a,v} &= \frac{1}{2}K_v H^2 \\ &= \left(\frac{1}{2}\right)\left(5\ \frac{\text{lbf}}{\text{ft}^2}\right)(18 \text{ ft})^2 \\ &= 810 \text{ lbf/ft of wall length} \end{aligned}$$

The horizontal component of the active resultant force is

$$\begin{aligned} R_{a,h} &= \frac{1}{2}K_h H^2 \\ &= \left(\frac{1}{2}\right)\left(30\ \frac{\text{lbf}}{\text{ft}^2}\right)(18 \text{ ft})^2 \\ &= 4860 \text{ lbf/ft of wall length} \end{aligned}$$

The resistance to sliding for a keyless foundation is

$$\begin{aligned} R_{\text{SL}} &= \left(\sum W_i + R_{a,v}\right)\tan\delta + c_A B \\ &= \left(\sum \gamma_i A_i + R_{a,v}\right)\tan\delta + c_A B \\ &= \left(\begin{array}{l} \left(145\ \frac{\text{lbf}}{\text{ft}^3}\right)\left(\begin{array}{l}(8 \text{ ft})(18 \text{ ft}) \\ \quad + (2)\left(\frac{1}{2}\right)(1 \text{ ft})(18 \text{ ft})\end{array}\right) \\ \quad + 810\ \frac{\text{lbf}}{\text{ft}} \end{array}\right) \\ &\quad \times\tan 18^\circ + \left(350\ \frac{\text{lbf}}{\text{ft}^2}\right)(10 \text{ ft}) \\ &= 11{,}396 \text{ lbf/ft of wall length} \end{aligned}$$

Calculate the factor of safety against sliding.

$$\begin{aligned} F_{\text{SL}} &= \frac{R_{\text{SL}}}{R_{a,h}} \\ &= \frac{11{,}396\ \frac{\text{lbf}}{\text{ft}}}{4860\ \frac{\text{lbf}}{\text{ft}}} \\ &= 2.3 \end{aligned}$$

Author Commentary

The earth pressure constant, K, incorporates an assumed soil density. K has units of lbf/ft^2 (lbf/ft^2 per linear foot of wall). Using K values saves time, but be careful to pay attention to units when doing so.

The answer is (C).

62. Determine the active earth pressure acting at the base of the wall using the Rankine earth pressure theory.

Calculate the coefficient of active earth pressure using the Rankine equation.

$$\begin{aligned} k_a &= \frac{1 - \sin\phi}{1 + \sin\phi} \\ &= \frac{1 - \sin 31^\circ}{1 + \sin 31^\circ} \\ &= 0.32 \end{aligned}$$

Calculate the active resultant force of the triangular pressure distribution.

$$\begin{aligned} R_a &= \frac{1}{2}p_a H = \frac{1}{2}k_a\gamma H^2 \\ &= \left(\frac{1}{2}\right)(0.32)\left(108.2\ \frac{\text{lbf}}{\text{ft}^3}\right)(15 \text{ ft})^2 \\ &= 3895 \text{ lbf/ft} \quad (3900 \text{ lbf/ft}) \end{aligned}$$

Author Commentary

The Rankine method assumes a straight wall face against the backfill and a horizontal backfill surface. When both of these conditions are met, the simplified Rankine equation may be used.

The answer is (C).

63. Use the Rankine earth pressure theory to find the active earth pressures and resultant forces shown in the illustration. Then, sum the moments in a counterclockwise direction about the toe (point A) to find the overturning moment.

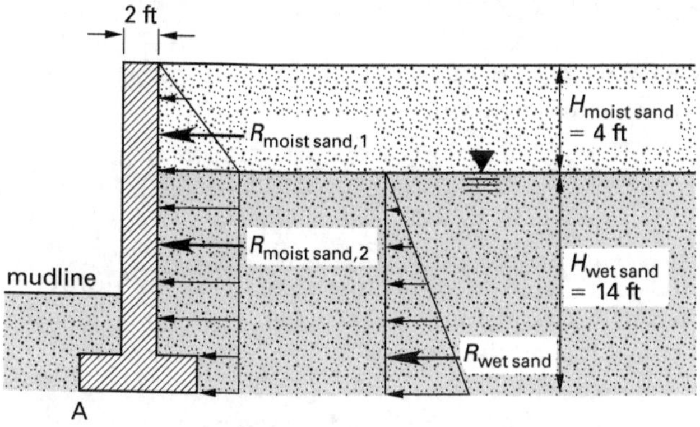

(not to scale)

Calculate the coefficient of active earth pressure using the Rankine equation.

$$k_a = \frac{1-\sin\phi}{1+\sin\phi} = \frac{1-\sin 26°}{1+\sin 26°} = 0.39$$

Calculate the horizontal active pressure at the base of the moist sand.

$$\begin{aligned} p_{a,h,\text{moist sand}} &= k_a\gamma_{\text{dry}}(1+w_{\text{moist sand}})H_{\text{moist sand}} \\ &= (0.39)\left(104.7\ \frac{\text{lbf}}{\text{ft}^3}\right)(1+0.03)(4\ \text{ft}) \\ &= 168\ \text{lbf/ft}^2 \end{aligned}$$

Find the horizontal active resultant at the base of the moist sand layer.

$$\begin{aligned} R_{a,h,\text{moist sand},1} &= \frac{1}{2}p_{a,h,\text{moist sand}}H_{\text{moist sand}} \\ &= \left(\frac{1}{2}\right)\left(168\ \frac{\text{lbf}}{\text{ft}^2}\right)(4\ \text{ft}) \\ &= 336\ \text{lbf/ft of wall width} \end{aligned}$$

Calculate the horizontal active resultant force that originates from the moist sand pressure that extends downward into the wet sand layer.

$$\begin{aligned} R_{a,h,\text{moist sand},2} &= p_{a,h,\text{moist sand}}H_{\text{wet sand}} \\ &= \left(168\ \frac{\text{lbf}}{\text{ft}^2}\right)(14\ \text{ft}) \\ &= 2352\ \text{lbf/ft of wall width} \end{aligned}$$

Calculate the horizontal active pressure at the base of the wet sand that acts at the base of the wall.

$$\begin{aligned} p_{a,h,\text{wet sand}} &= k_a\gamma_{\text{dry}}(1+w_{\text{wet sand}})H_{\text{wet sand}} \\ &= (0.39)\left(104.7\ \frac{\text{lbf}}{\text{ft}^3}\right)(1+0.13)(14\ \text{ft}) \\ &= 646\ \text{lbf/ft}^2 \end{aligned}$$

The active horizontal resultant at the base of the wet sand is

$$\begin{aligned} R_{a,h,\text{wet sand}} &= \frac{1}{2}p_{a,h,\text{wet sand}}H_{\text{wet sand}} \\ &= \left(\frac{1}{2}\right)\left(646\ \frac{\text{lbf}}{\text{ft}^2}\right)(14\ \text{ft}) \\ &= 4522\ \text{lbf/ft of wall width} \end{aligned}$$

Use a counterclockwise summation about the toe (point A) to determine the overturning moment.

$$\begin{aligned} \sum M_{\text{A,OT}} &= \frac{1}{3}H_{\text{wet sand}}R_{a,h,\text{wet sand}} \\ &\quad +\frac{1}{2}H_{\text{wet sand}}R_{a,h,\text{moist sand},2} \\ &\quad +\left(\frac{1}{3}H_{\text{moist sand}}+H_{\text{wet sand}}\right)R_{a,h,\text{moist sand},1} \\ &= \left(\frac{1}{3}\right)(14\ \text{ft})\left(4522\ \frac{\text{lbf}}{\text{ft}}\right)+\left(\frac{1}{2}\right)(14\ \text{ft})\left(2352\ \frac{\text{lbf}}{\text{ft}}\right) \\ &\quad +\left(\left(\frac{1}{3}\right)(4\ \text{ft})+14\ \text{ft}\right)\left(336\ \frac{\text{lbf}}{\text{ft}}\right) \\ &= 42{,}719\ \text{ft-lbf/ft of wall width} \end{aligned}$$

To calculate the resisting moment about the toe of the retaining wall, determine the self-weight and soil weights above the heel. Divide the wall and soil into three centroids composed of a vertical rectangle (section 1), a horizontal rectangle (section 2), and another vertical rectangle for the moist and wet sands (section 3).

The weight of section 1 is

$$\begin{aligned} W_{\text{sec 1}} &= t_{\text{wall}}(H_{\text{moist sand}}+H_{\text{wet sand}}-t_{\text{base}})\gamma_{\text{concrete}} \\ &= (2\ \text{ft})(4\ \text{ft}+14\ \text{ft}-2\ \text{ft})\left(145\ \frac{\text{lbf}}{\text{ft}^3}\right) \\ &= 4640\ \text{lbf/ft of wall width} \end{aligned}$$

The weight of section 2 is

$$\begin{aligned} W_{\text{sec}\,2} &= Bt_{\text{base}}\gamma_{\text{concrete}} \\ &= (4\ \text{ft})(2\ \text{ft})\left(145\ \frac{\text{lbf}}{\text{ft}^3}\right) \\ &= 1160\ \text{lbf/ft of wall width} \end{aligned}$$

The weight of section 3 is

$$\begin{aligned} W_{\text{sec}\,3} &= \left(B - \frac{B}{2} - \frac{t_{\text{wall}}}{2}\right)H_{\text{moist sand}}\gamma_{\text{dry}}(1 + w_{\text{moist sand}}) \\ &\quad + \left(B - \frac{B}{2} - \frac{t_{\text{wall}}}{2}\right)(H_{\text{wet sand}} - t_{\text{base}}) \\ &\quad \times \gamma_{\text{dry}}(1 + w_{\text{wet sand}}) \\ &= \left(4\ \text{ft} - \frac{4\ \text{ft}}{2} - \frac{2\ \text{ft}}{2}\right)(4\ \text{ft})\left(104.7\ \frac{\text{lbf}}{\text{ft}^3}\right)(1 + 0.03) \\ &\quad + \left(4\ \text{ft} - \frac{4\ \text{ft}}{2} - \frac{2\ \text{ft}}{2}\right)(14\ \text{ft} - 2\ \text{ft}) \\ &\quad \times \left(104.7\ \frac{\text{lbf}}{\text{ft}^3}\right)(1 + 0.13) \\ &= 1851\ \text{lbf/ft of wall width} \end{aligned}$$

Sum the resisting moment about the toe (point A), considering clockwise to be the positive direction.

$$\begin{aligned} \sum M_{\text{A},R} &= \left(\frac{B}{2}\right)W_{\text{sec}\,1} + \left(\frac{B}{2}\right)W_{\text{sec}\,2} + \frac{7}{8}BW_{\text{sec}\,3} \\ &= \left(\frac{4\ \text{ft}}{2}\right)\left(4640\ \frac{\text{lbf}}{\text{ft}}\right) + \left(\frac{4\ \text{ft}}{2}\right)\left(1160\ \frac{\text{lbf}}{\text{ft}}\right) \\ &\quad + \left(\frac{7}{8}\right)(4\ \text{ft})\left(1851\ \frac{\text{lbf}}{\text{ft}}\right) \\ &= 18{,}079\ \text{ft-lbf/ft of wall width} \end{aligned}$$

Find the factor of safety against overturning.

$$\begin{aligned} F_{\text{OT}} &= \frac{M_{\text{resisting,toe}}}{M_{\text{overturning,toe}}} = \frac{M_{\text{A},R}}{M_{\text{A,OT}}} \\ &= \frac{18{,}079\ \frac{\text{ft-lbf}}{\text{ft}}}{42{,}719\ \frac{\text{ft-lbf}}{\text{ft}}} \\ &= 0.42 \end{aligned}$$

Author Commentary

Sketching the wall and backfill for quick reference during the solution will save time on the exam.

The answer is (B).

64. The development of tension in rod 6 is due to active earth pressure, and can be calculated per foot of wall width. Use the area of the lateral and vertical spacing of the rods to determine the tension.

$$T_i = \gamma d_i hsk_a = (\gamma_{\text{dry}}(1 + w))d_i s_h s_v k_a$$

Calculate the coefficient of active earth pressure using the Rankine equation.

$$k_a = \frac{1 - \sin\phi}{1 + \sin\phi} = \frac{1 - \sin 27°}{1 + \sin 27°} = 0.38$$

Rod 6, T_6, is located at a depth of 1.5 ft + (5)(3 ft) = 16.5 ft. The horizontal rod spacing, s_h, is 2.5 ft, and the vertical rod spacing, s_v, is 3 ft. The tension in rod 6 is

$$\begin{aligned} T_6 &= (\gamma_{\text{dry}}(1 + w))d_6 s_h s_v k_a \\ &= \frac{\left(\left(106.2\ \frac{\text{lbf}}{\text{ft}^3}\right)(1 + 0.14)\right) \times (16.5\ \text{ft})(2.5\ \text{ft})(3\ \text{ft})(0.38)}{1000\ \frac{\text{lbf}}{\text{kip}}} \\ &= 5.7\ \text{kips} \end{aligned}$$

Author Commentary

Be sure to recognize which values are needed for the solution and which are not. The 30 ft excavation depth shown in the illustration is not necessary to solve the problem.

The answer is (B).

65. For calculation purposes, assume the depth of the excavation is larger than the width. The surface surcharge loading, q, is zero. Rearrange the factor of safety equation to determine the bearing capacity factor, N_c.

$$F = \frac{N_c c}{\gamma H + q}$$

$$\begin{aligned} N_c &= \frac{F(\gamma H + q)}{c} = \frac{F\left(\gamma_{\text{dry}}(1 + w)H + q\right)}{c} \\ &= \frac{(1.3)\left(\left(101.2\ \frac{\text{lbf}}{\text{ft}^3}\right)(1 + 0.143)(30\ \text{ft}) + 0\ \frac{\text{lbf}}{\text{ft}^2}\right)}{620\ \frac{\text{lbf}}{\text{ft}^2}} \\ &= 7.28 \end{aligned}$$

Confirm that the critical height is greater than the excavation depth.

$$\begin{aligned}H_c &= \frac{N_c c}{\gamma} = \frac{N_c c}{\gamma_{\text{dry}}(1+w)}\\ &= \frac{(7.28)\left(620\ \frac{\text{lbf}}{\text{ft}^2}\right)}{\left(101.2\ \frac{\text{lbf}}{\text{ft}^3}\right)(1+0.143)}\\ &= 39\ \text{ft}\quad [>30\ \text{ft, so OK}]\end{aligned}$$

For $N_c = 7.3$, the maximum allowable width of the excavation is

$$\begin{aligned}\frac{H}{B} &= 3\\ B &= \frac{H}{3} = \frac{30\ \text{ft}}{3}\\ &= 10\ \text{ft}\end{aligned}$$

Author Commentary

Assuming that the depth of the excavation will be larger than the width allows the use of the basic factor of safety equation, so the problem may be solved more quickly.

The answer is (B).

66. Calculate the horizontal active earth pressure at the cut line using the coefficient of active earth pressure. The coefficient of active earth pressure is

$$k_a = \frac{1-\sin\phi}{1+\sin\phi} = \frac{1-\sin 25^\circ}{1+\sin 25^\circ} = 0.41$$

The horizontal active earth pressure at the cut line is

$$\begin{aligned}p_{a,h,\text{cut}} &= p_v k_a = (\gamma_{\text{dry}}(1+w)H)k_a\\ &= \left(\left(103.1\ \frac{\text{lbf}}{\text{ft}^3}\right)(1+0.07)(25\ \text{ft})\right)(0.41)\\ &= 1131\ \text{lbf/ft}^2\end{aligned}$$

The coefficient of passive earth pressure is

$$\begin{aligned}k_p &= \frac{1+\sin\phi}{1-\sin\phi} = \frac{1+\sin 25^\circ}{1-\sin 25^\circ}\\ &= 2.46\end{aligned}$$

Calculate the net horizontal passive earth pressure at the base of the bulkhead by finding the difference between the total passive earth pressure and the total active earth pressure. The depth of installation, D, for the bulkhead below the proposed excavation depth is 15 ft. Use this value to determine the total passive earth pressure.

$$\begin{aligned}p_{p,h,\text{bulkhead base}} &= k_p(\gamma_{\text{dry}}(1+w)D)\\ &\quad -k_a\gamma_{\text{dry}}(1+w)(H+D)\\ &= (2.46)\left(103.1\ \frac{\text{lbf}}{\text{ft}^3}\right)\\ &\quad \times(1+0.07)(15\ \text{ft})\\ &\quad -(0.41)\left(103.1\ \frac{\text{lbf}}{\text{ft}^3}\right)\\ &\quad \times(1+0.07)(25\ \text{ft}+15\ \text{ft})\\ &= 2268\ \text{lbf/ft}^2\end{aligned}$$

Through static equilibrium, the pivot point can be calculated by comparing areas of stress distribution. This is represented by the equation for determining the pivot point height, H_p, above the bottom tip of the bulkhead as shown.

$$\begin{aligned}\frac{1}{2}p_{a,h,\text{cut}}(D-H_p) &= \frac{1}{2}p_{a,h,\text{bulkhead base}}H_p\\ H_p &= \frac{p_{a,h,\text{cut}}D}{p_{a,h,\text{cut}}+p_{a,h,\text{bulkhead base}}}\\ &= \frac{\left(1131\ \frac{\text{lbf}}{\text{ft}^2}\right)(40\ \text{ft}-25\ \text{ft})}{1131\ \frac{\text{lbf}}{\text{ft}^2}+2268\ \frac{\text{lbf}}{\text{ft}^2}}\\ &= 5\ \text{ft}\end{aligned}$$

Perform a static moment analysis to determine the tensile force required in the tie rod, ignoring the passive pressure and using the clockwise direction about the pivot point as positive.

$$\sum M_{\text{pivot point}} = T_{\text{rod}}L_{\text{rod}} - R_{a,\text{sand}}L_{\text{sand}} = 0$$

$$\begin{aligned}T_{\text{rod}} &= \frac{R_{a,\text{sand}}\left(\frac{1}{3}H+D-H_p\right)}{H+D-D_{\text{tie rod}}-H_p}\\ &= \frac{\frac{1}{2}p_{a,h,\text{cut}}H\left(\frac{1}{3}H+D-H_p\right)}{H+D-D_{\text{tie rod}}-H_p}\\ &= \frac{\left(\frac{1}{2}\right)\left(1131\ \frac{\text{lbf}}{\text{ft}^2}\right)(25\ \text{ft})\left(\left(\frac{1}{3}\right)(25\ \text{ft})+15\ \text{ft}-5\ \text{ft}\right)}{25\ \text{ft}+15\ \text{ft}-6\ \text{ft}-5\ \text{ft}}\\ &= 8938\ \text{lbf/ft of bulkhead length}\quad (8900\ \text{lbf/ft})\end{aligned}$$

Author Commentary

The pivot point of the bulkhead coincides with the location where passive pressure transitions to active pressure. Similar to structural evaluations, using static equilibrium to compare the active and passive stress distribution areas is a common method of determining the location.

The answer is (C).

67. Since struts are placed at every 6 ft of depth, a strut is present at the midpoint depth of the excavation, 18 ft. Use the Tschebotarioff trapezoidal earth pressure distribution equation.

$$p_{\text{max}} = 0.8k_a\gamma H = 0.8k_a\gamma_{\text{dry}}(1+w)H$$

Calculate the coefficient of active earth pressure.

$$k_a = \frac{1-\sin\phi}{1+\sin\phi} = \frac{1-\sin 33^\circ}{1+\sin 33^\circ} = 0.29$$

The maximum horizontal earth pressure of the bracing at the midpoint of the excavation is

$$\begin{aligned} p_{\text{max}} &= 0.8k_a\gamma_{\text{dry}}(1+w)H \\ &= (0.8)(0.29)\left(107.7\ \frac{\text{lbf}}{\text{ft}^3}\right)(1+0.04)(36\ \text{ft}) \\ &= 935.7\ \text{lbf/ft}^2 \end{aligned}$$

The contributory area for the strut, A_c, is the depth per foot of bracing support width. Calculate the load on the strut per foot of bracing width.

$$\begin{aligned} R_{\text{max at 18 ft}} &= p_{\text{max}}A_c \\ &= \left(935.7\ \frac{\text{lbf}}{\text{ft}^2}\right)(6\ \text{ft}) \\ &= 5614\ \text{lbf/ft bracing length} \quad (5600\ \text{lbf/ft}) \end{aligned}$$

Author Commentary

Since dry unit weight is given in the problem statement, it must be converted to total unit weight during the calculation of p_{max}.

The answer is (C).

68. The development of tension in each strip can be calculated per foot of wall width using the coefficient of active earth pressure. Use the area of the horizontal spacing, s_h, and vertical spacing, s_v, of the strips to determine the tension value for each strip.

$$T_i = \gamma d_i s_h s_v k_a$$

Determine the coefficient of active earth pressure using the Rankine equation.

$$k_a = \frac{1-\sin\phi}{1+\sin\phi} = \frac{1-\sin 25^\circ}{1+\sin 25^\circ} = 0.41$$

Calculate the tension in strip 1, T_1.

$$\begin{aligned} T_1 &= \gamma d_1 s_h s_v k_a \\ &= \left(116.5\ \frac{\text{lbf}}{\text{ft}^3}\right)(1.25\ \text{ft})(3\ \text{ft})(2.5\ \text{ft})(0.41) \\ &= 448\ \text{lbf} \end{aligned}$$

The depth to each strip is based on the given vertical spacing. The depth to the first strip is half the vertical spacing.

Find and tabulate the tension values for the other strips.

strip	depth to strip, d_i (ft)	tension, T_i (lbf)
1	1.25	448
2	3.75	1343
3	6.25	2239
4	8.75	3135
5	11.25	4030
6	13.75	4926
7	16.25	5821
8	18.75	6717
9	21.25	7613
10	23.75	8508
		$\sum T_i = 44{,}780$

Use the Rankine earth pressure theory to find the active earth pressure resultant force that acts on the wall from the granular soil.

Calculate the resultant of the triangular pressure distribution.

$$\begin{aligned} R_{a,h} &= \frac{1}{2}p_a H = \frac{1}{2}k_a\gamma H^2 \\ &= \left(\frac{1}{2}\right)(0.41)\left(116.5\ \frac{\text{lbf}}{\text{ft}^3}\right)(25\ \text{ft})^2 \\ &= 14{,}927\ \text{lbf/ft of wall width} \end{aligned}$$

The combined tension force from all strips divided by the horizontal 3 ft strip spacing must be equivalent to the horizontal component of the active earth pressure resultant force.

$$\frac{\sum T_i}{h} = \frac{44{,}780 \text{ lbf}}{3 \text{ ft}} = 14{,}927 \text{ lbf/ft} \quad [= R_{a,h}, \text{ so OK}]$$

The total length of each strip is the sum of the effective length that extends beyond the Rankine active zone (i.e., area above the failure plane) and the length within the Rankine active zone.

$$L_t = L_e + L_R$$

The frictional resistance develops on both sides of the strip. The effective length is twice the strip width, and the minimum effective length must occur at the uppermost strip. Therefore, the tension and depth combination for the uppermost strip should be selected.

$$\begin{aligned} T_i &= 2bL_e f\gamma d_i \\ &= 2bL_e(\tan\delta)\gamma d_i \\ L_e &= \frac{T_i}{2b(\tan\delta)\gamma d_i} \\ &= \frac{T_i}{(2)\left(\frac{2.5 \text{ in}}{12 \frac{\text{in}}{\text{ft}}}\right)(\tan 18°)\left(116.5 \frac{\text{lbf}}{\text{ft}^3}\right)d_i} \\ &= \frac{T_i}{\left(15.8 \frac{\text{lbf}}{\text{ft}^2}\right)d_i} \end{aligned}$$

The effective length for strip 1 is

$$\begin{aligned} L_{e,1} &= \frac{T_1}{\left(15.8 \frac{\text{lbf}}{\text{ft}^2}\right)d_1} \\ &= \frac{448 \text{ lbf}}{\left(15.8 \frac{\text{lbf}}{\text{ft}^2}\right)(1.25 \text{ ft})} \\ &= 22.7 \text{ ft} \end{aligned}$$

Check the effective length of another strip to be sure it is consistent. The effective length for strip 10 is

$$\begin{aligned} L_{e,10} &= \frac{T_{10}}{\left(15.8 \frac{\text{lbf}}{\text{ft}^2}\right)d_{10}} \\ &= \frac{8508 \text{ lbf}}{\left(15.8 \frac{\text{lbf}}{\text{ft}^2}\right)(23.75 \text{ ft})} \\ &= 22.7 \text{ ft} \end{aligned}$$

Determine the strip length within the Rankine active zone by finding the angle of the failure plane with the horizontal (parallel with the level backfill surface). Equate the width of the active zone at the widest part, near the top, with the length of the strip in the Rankine active zone, L_R.

$$\begin{aligned} \theta &= 45° + \frac{\phi}{2} = 45° + \frac{25°}{2} \\ &= 57.5° \end{aligned}$$

The length of the strip in the Rankine active zone is

$$\begin{aligned} L_R &= H \tan(90° - \theta) \\ &= (25 \text{ ft})\tan(90° - 57.5°) \\ &= 15.9 \text{ ft} \end{aligned}$$

Calculate the total strip length.

$$\begin{aligned} L_t &= L_e + L_R \\ &= 22.7 \text{ ft} + 15.9 \text{ ft} \\ &= 38.6 \text{ ft} \quad (39 \text{ ft}) \end{aligned}$$

Author Commentary

Using a table to keep track of calculations may help save time in problems where repetitive calculations are required.

The answer is (D).

69. An aquifer is a geological formation that can hold water. A portion of the aquifer may or may not be saturated. Based on the location of the impermeable clay layer, the aquifer thickness for the soil profile shown is 100 ft + 100 ft = 200 ft.

Author Commentary

An aquifer is identified based on the presence of porous geological material (e.g., sand and gravel), not whether water is present. If the material has the ability to store water, the formation is considered to be an aquifer.

The answer is (B).

70. Based on the water table location, the aquifer is divided into two zones. The vadose zone is located above the water table elevation, and can be either saturated, partially saturated, or empty. The phreatic zone is located below the water table, where the soils are submerged. A confined zone is a layer bounded by impermeable material on both sides. An artesian zone is where water under pressure rises up above such a confining layer.

Layer 2 is submerged below the water table, and is therefore part of the phreatic zone.

The answer is (C).

71. The effective velocity can be calculated from the overall groundwater flow velocity and the porosity.

$$\mathrm{v}_e = n\mathrm{v}_{\text{pore}}$$

$$\mathrm{v}_{\text{pore}} = \frac{\mathrm{v}_e}{n} = \frac{\left(\frac{3\ \text{mi}}{180\ \text{days}}\right)\left(5280\ \frac{\text{ft}}{\text{mi}}\right)\left(12\ \frac{\text{in}}{\text{ft}}\right)}{(0.12)\left(86{,}400\ \frac{\text{sec}}{\text{day}}\right)} = 0.10\ \text{in/sec}$$

Author Commentary

An aquifer can be divided into zones depending on the water table location. In this problem, layers 1 and 2 could be considered vadose and saturated zones, depending on the moisture conditions. However, the soil type, solids density, and porosity remain the same throughout both zones.

The answer is (C).

72. The groundwater flow rate can be calculated from the aquifer width, transmissivity, and hydraulic gradient.

$$Q = bT_i = bT\left(\frac{\Delta H}{L}\right)$$

Determine the transmissivity using the hydraulic conductivity, K, and original aquifer phreatic zone thickness, Y.

$$T = KY = \left(\frac{1000\ \frac{\text{gal}}{\text{day-ft}^2}}{7.48\ \frac{\text{gal}}{\text{ft}^3}}\right)(100\ \text{ft}) = 13{,}369\ \text{ft}^3/\text{day-ft}$$

Calculate the overall groundwater flow rate.

$$Q = bT\left(\frac{\Delta H}{L}\right) = \frac{(50\ \text{mi})\left(13{,}369\ \frac{\text{ft}^3}{\text{day-ft}}\right)\left(\frac{200\ \text{ft}}{100\ \text{mi}}\right)}{86{,}400\ \frac{\text{sec}}{\text{day}}} = 15.47\ \text{ft}^3/\text{sec}\quad (15\ \text{ft}^3/\text{sec})$$

The answer is (B).

73. Since the proposed well penetrates the full depth of the aquifer down to the impermeable layer, the Dupuit equation can be used to determine the well's radius of influence.

$$Q = \frac{\pi K(y_1^2 - y_2^2)}{\ln\frac{r_1}{r_2}}$$

The aquifer thickness is the difference between the depth to impermeable rock and the aquifer phreatic surface.

$$Y = 500\ \text{ft} - 100\ \text{ft} = 400\ \text{ft}$$

Calculate the hydraulic conductivity of the aquifer.

$$T = KY$$

$$K = \frac{T}{Y} = \frac{8000\ \frac{\text{gal}}{\text{day-ft}}}{400\ \text{ft}} = 20\ \text{gal/day-ft}^2$$

The drawdown occurs at the top of the screen, 200 ft below ground surface. From the Dupuit equation, the radius of influence is

$$Q = \frac{\pi K(y_1^2 - y_2^2)}{\ln\frac{r_1}{r_2}}$$

$$r_1 = r_2 e^{\frac{\pi K(y_1^2 - y_2^2)}{Q}} = \left(\frac{\frac{18\ \text{in}}{2}}{12\ \frac{\text{in}}{\text{ft}}}\right)\left(e^{\frac{\pi\left(20\ \frac{\text{gal}}{\text{day-ft}^2}\right)\left((500\ \text{ft}-100\ \text{ft})^2-(500\ \text{ft}-200\ \text{ft})^2\right)}{\left(500\ \frac{\text{gal}}{\text{min}}\right)\left(60\ \frac{\text{min}}{\text{hr}}\right)\left(24\ \frac{\text{hr}}{\text{day}}\right)}}\right) = 338\ \text{ft}$$

Determine the minimum distance the new well can be constructed from the existing well.

$$\begin{aligned} D_{\text{req}} &= r_{\text{existing well}} + d_{\text{req spacing}} + r_1 \\ &= 700 \text{ ft} + 500 \text{ ft} + 338 \text{ ft} \\ &= 1538 \text{ ft} \quad (1500 \text{ ft}) \end{aligned}$$

Author Commentary

Be aware that well pipe sizes are typically given according to diameter, not radius. Be certain to use the pipe radius in the Dupuit equation.

The answer is (C).

74. Since the well is in artesian conditions (i.e., the water is under pressure and rising above the bottom of the low permeability zone) and is fed by a confined aquifer, the safe yield can be found using the Thiem equation. The safe yield is the flow rate that can be obtained without exceeding the given criteria.

$$Q = \frac{2\pi KY(y_1 - y_2)}{\ln \frac{r_1}{r_2}}$$

The aquifer thickness is the difference between the depth of the well penetration and thickness of the low permeability zone.

$$Y = 200 \text{ ft} - 125 \text{ ft} = 75 \text{ ft}$$

The maximum radius of influence allowed is equal to the observed water level in the well when water is not being pumped.

$$r_1 = 1000 \text{ ft}$$

The maximum drawdown allowed during pumping should coincide with the well radius.

$$r_2 = \frac{\frac{20 \text{ in}}{2}}{12 \frac{\text{in}}{\text{ft}}} = 0.83 \text{ ft}$$

The aquifer thickness at the radial distance r_1 is equivalent to the water level that rises in the well.

$$y_1 = 200 \text{ ft} - 75 \text{ ft} = 125 \text{ ft}$$

The aquifer depth at the radial distance r_2 is

$$y_2 = 125 \text{ ft} - 30 \text{ ft} = 95 \text{ ft}$$

The safe yield is

$$\begin{aligned} Q &= \frac{2\pi KY(y_1 - y_2)}{\ln \frac{r_1}{r_2}} \\ &= \frac{2\pi \left(100 \frac{\text{gal}}{\text{day-ft}^2}\right)(75 \text{ ft})(125 \text{ ft} - 95 \text{ ft})}{\ln \frac{1000 \text{ ft}}{0.83 \text{ ft}}} \\ &= 199{,}281 \text{ gal/day} \quad (200{,}000 \text{ gal/day}) \end{aligned}$$

Author Commentary

The difference between y_1 and y_2 is the drawdown, which is the same as the difference between aquifer thicknesses in artesian conditions. To save time, use 30 ft as the drawdown instead of going through all the calculations shown in this solution.

The answer is (B).

75. Use Darcy's law to find the initial flow rate. The hydraulic gradient, i, can be calculated from the water depth above ground level and the thickness of the subgrade soils above the well points.

$$i = \frac{H}{t} = \frac{4 \text{ ft}}{12 \text{ ft} - 4 \text{ ft}} = 0.5 \text{ ft/ft}$$

The initial flow rate is

$$\begin{aligned} Q &= KiA \\ &= \frac{\left(5 \times 10^{-4} \frac{\text{in}}{\text{sec}}\right)\left(0.5 \frac{\text{ft}}{\text{ft}}\right)(100{,}000 \text{ ft}^2)}{12 \frac{\text{in}}{\text{ft}}} \\ &= 2.08 \text{ ft}^3\text{/sec} \quad (2.1 \text{ ft}^3\text{/sec}) \end{aligned}$$

Author Commentary

Since the flow is one-dimensional, free gravitational flow, Darcy's law may be used to determine the initial flow rate. Keep in mind that as the water elevation is reduced, the flow rate through the soil will decrease with the hydraulic gradient.

The answer is (C).

76. Use the Horton equation to determine the cumulative infiltration that occurs during the 2 hr rainfall event.

$$\begin{aligned} F_t &= f_c t + \left(\frac{f_0 - f_c}{k}\right)(1 - e^{-kt}) \\ &= \left(0.16\ \frac{\text{in}}{\text{hr}}\right)(2\ \text{hr}) + \left(\frac{1.8\ \frac{\text{in}}{\text{hr}} - 0.16\ \frac{\text{in}}{\text{hr}}}{\frac{0.6}{\text{hr}}}\right) \\ &\quad \times \left(1 - e^{-\left(\frac{0.6}{\text{hr}}\right)(2\ \text{hr})}\right) \\ &= 2.23\ \text{in} \end{aligned}$$

Determine the average infiltration rate.

$$\begin{aligned} \bar{f} &= \frac{F_t}{t_{\text{infiltration}}} = \frac{2.23\ \text{in}}{\frac{2\ \text{hr}}{3600\ \frac{\text{sec}}{\text{hr}}}} \\ &= 3.1 \times 10^{-4}\ \text{in/sec} \end{aligned}$$

Find the difference between the average infiltration rate and the hydraulic conductivity.

$$\begin{aligned} f_{\text{diff}} &= \bar{f} - K \\ &= 3.1 \times 10^{-4}\ \frac{\text{in}}{\text{sec}} - 2.5 \times 10^{-4}\ \frac{\text{in}}{\text{sec}} \\ &= 0.60 \times 10^{-4}\ \text{in/sec} \end{aligned}$$

Author Commentary

The Horton equation is used primarily in surface water hydrology. However, it also applies to some groundwater hydrology and geotechnical problems. Review the Horton equation to save time during the exam.

The answer is (A).

77. Equate the pullout capacity to the factor of safety and tensile strength.

$$\begin{aligned} \text{FS} &= \frac{P_u}{T} \\ P_u &= (\text{FS})T \end{aligned}$$

The pullout capacity in terms of the surface area of the rock bolt is

$$\begin{aligned} P_u &= Ab_r \\ &= (2\pi rL)b_r \end{aligned}$$

Combine the equations, and solve for the bond strength.

$$\begin{aligned} b_r &= \frac{P_u}{A} \\ &= \frac{(\text{FS})T}{2\pi rL} \\ &= \frac{(3)(200\ \text{kips})\left(1000\ \frac{\text{lbf}}{\text{kip}}\right)}{2\pi\left(\frac{4\ \text{in}}{2}\right)(8\ \text{ft})\left(12\ \frac{\text{in}}{\text{ft}}\right)} \\ &= 497\ \text{lbf/in}^2 \quad (500\ \text{lbf/in}^2) \end{aligned}$$

Author Commentary

Be sure the dimensions for the bolt and embedment length are correct.

The answer is (D).

78. Use the equation for maximum lateral pressure of a braced cut in drained sand to determine the allowable depth of the excavation. The depth of the excavation is equivalent to the height of the cut.

$$p_{\max} = 0.65k_a\gamma H$$

Since the pressure distribution on the horizontal lagging is rectangular, use the equation for maximum moment in a simply supported beam. The maximum allowable pressure occurs at the center of the beam and can be used to determine the allowable excavation depth. Solve for H to determine the allowable depth per foot of lagging width.

$$\begin{aligned} M &= \frac{p_{\max}L^2}{8} = \frac{0.65k_a\gamma HL^2}{8} \\ H &= \frac{8M}{0.65k_a\gamma L^2} = \frac{8M}{(0.65)\left(\frac{1-\sin\phi}{1+\sin\phi}\right)\gamma L^2} \\ &= \frac{(8)(3600\ \text{ft-lbf})}{(0.65)\left(\frac{1-\sin 36^\circ}{1+\sin 36^\circ}\right)\left(111.8\ \frac{\text{lbf}}{\text{ft}^3}\right)(10\ \text{ft})^2} \\ &= 15.3\ \text{ft} \quad (15\ \text{ft}) \end{aligned}$$

Author Commentary

Tab beam equations in your reference material for use in shoring and tieback problems.

The answer is (B).

79. From OSHA 29 *Code of Federal Regulations* (CFR) 1926, Subpart P, App. A, the allowable bearing capacity for OSHA type A soil is 1.5 tons/ft^2. Determine the contact pressure for a single outrigger pad.

$$q_{\text{actual}} = \frac{P}{A} = \frac{100 \text{ tons}}{81 \text{ ft}^2} = 1.23 \text{ tons/ft}^2$$

The corresponding factor of safety is

$$\text{FS} = \frac{q_{\text{allowable}}}{q_{\text{actual}}} = \frac{1.5 \, \frac{\text{tons}}{\text{ft}^2}}{1.23 \, \frac{\text{tons}}{\text{ft}^2}} = 1.22 \quad (1.2)$$

Author Commentary

Be familiar with OSHA soil types, as they are typically used to establish allowable bearing capacity.

The answer is (C).

80. According to OSHA 29 *Code of Federal Regulations* (CFR) 1926.652, App. B, slopes are to be reduced by 50% if the soil shows signs of distress. The new slope is

$$S = \frac{1.5}{0.50} = 3 \quad (3{:}1)$$

Author Commentary

Tab frequently referenced charts and material in OSHA 1926 for easy access during the exam.

The answer is (D).

Grown Woman

LIFE FINANCIAL GOAL PLANNER

UNAPOLOGETICAL

Wisdom

GROWING TOGETHER IN WISDOM

GET OUT OF DEBT | SAVE MONEY
TAKE CONTROL OF YOUR FINANCES
12 MONTH UNDATED PLANNER